For Richard Banks Sutton
and William Banks Sutton

with love, from
Grandpa.

July 1978

The Prehistoric World

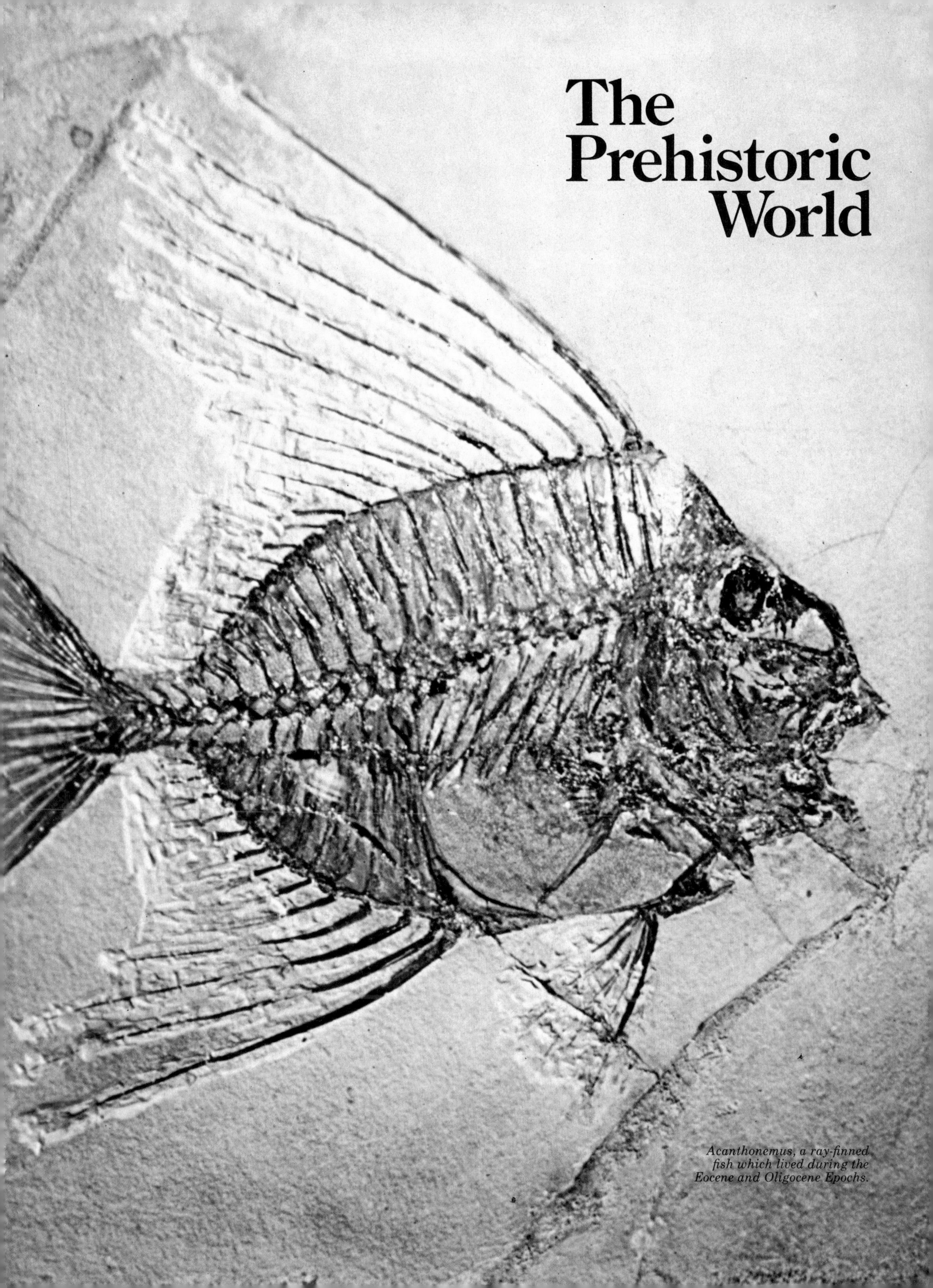

Acanthonemus, a ray-finned fish which lived during the Eocene and Oligocene Epochs.

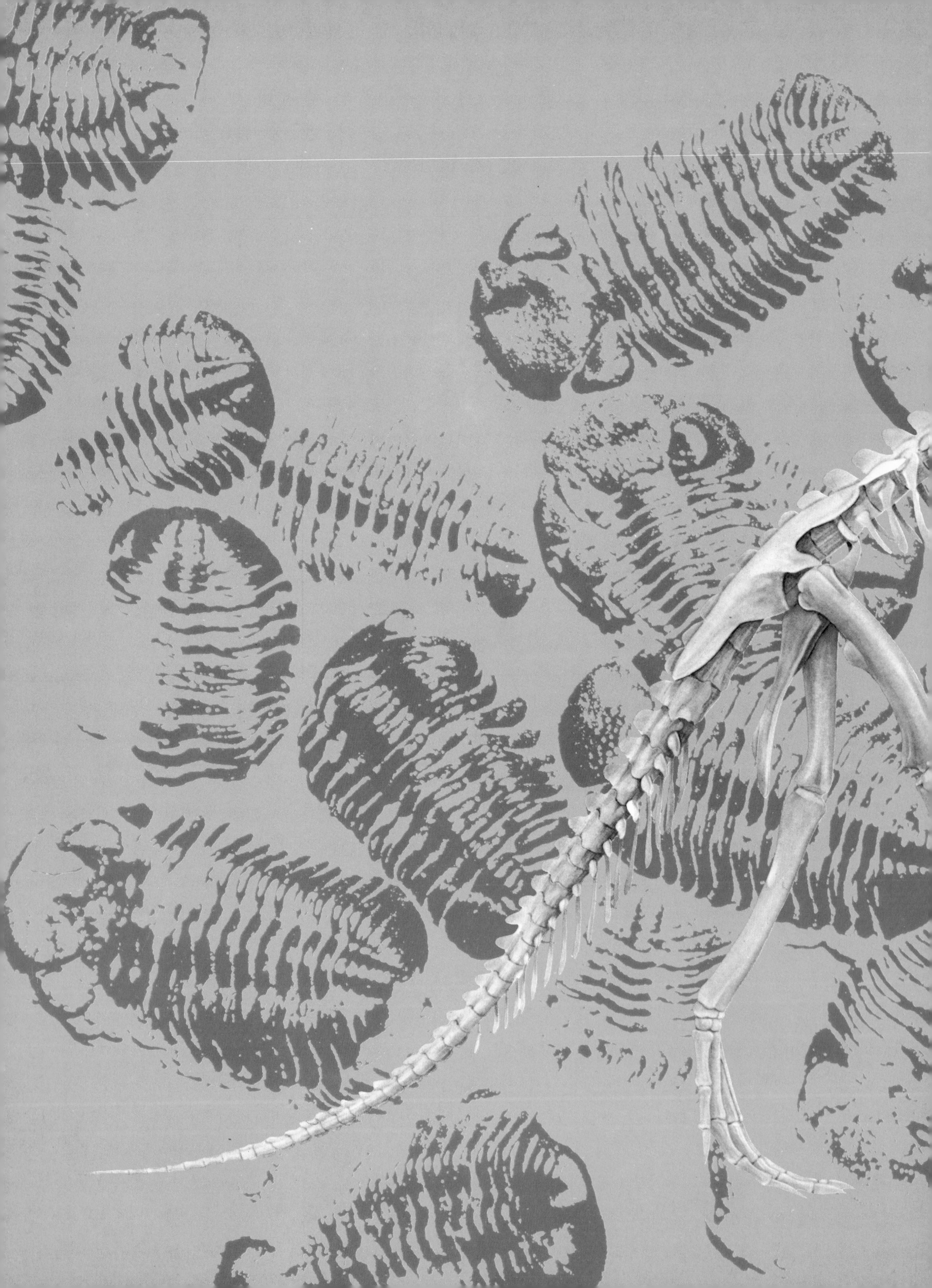

The Prehistoric World

WARWICK PRESS

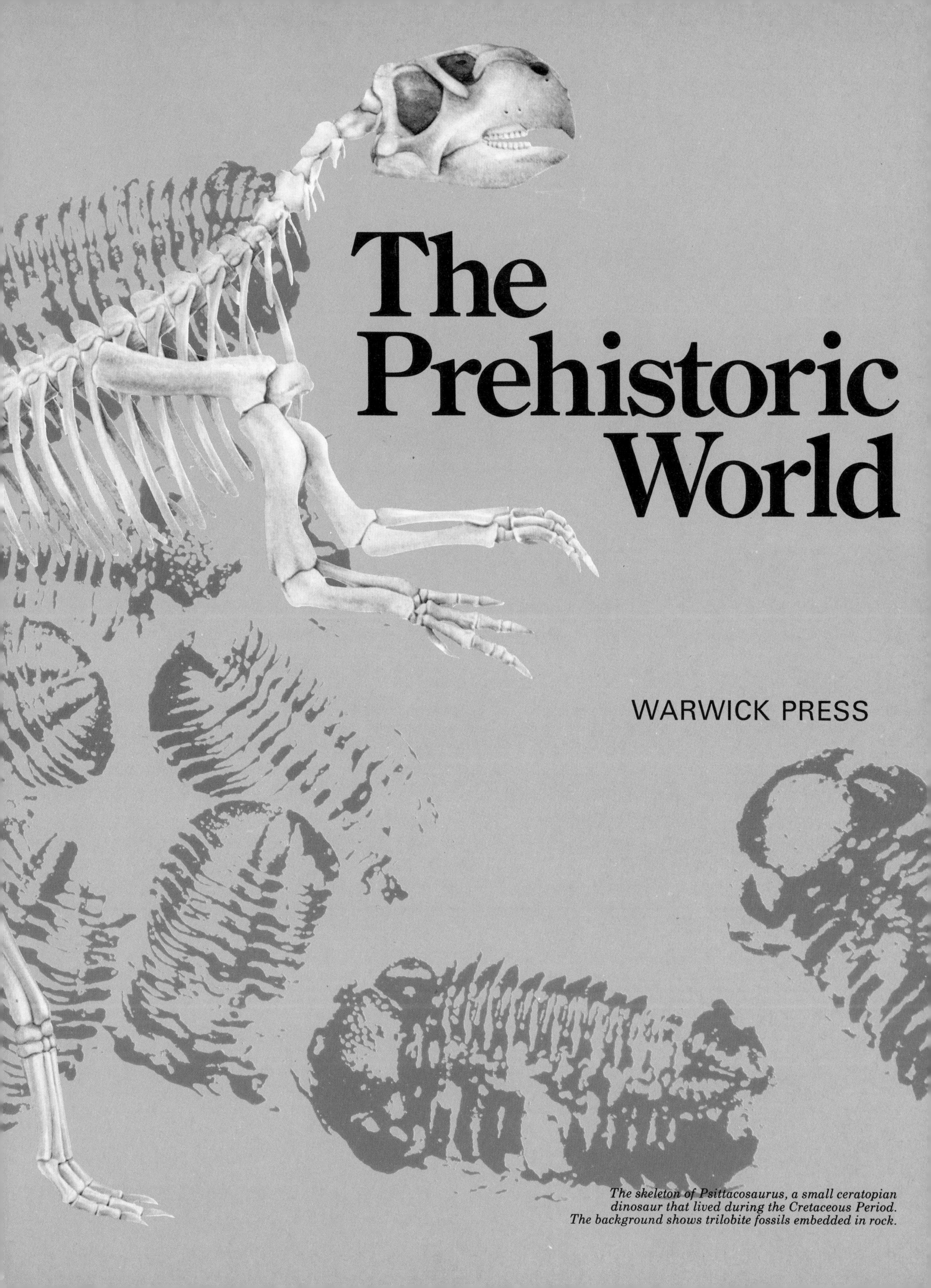

The skeleton of Psittacosaurus, a small ceratopian dinosaur that lived during the Cretaceous Period. The background shows trilobite fossils embedded in rock.

The fossilized leaf of a water lily shows the small size of the ray-finned fish Priscacaria.

Contributors

Alan Bartram M.A., Ph.D.
Basil Booth B.Sc., Ph.D., F.G.S.
Michael Chinery M.A.
Euan N. K. Clarkson B.A., Ph.D.
Barry Cox M.A., Ph.D., D.Sc.
Diane Edwards M.A., Ph.D.
Christopher Maynard B.A.
W. D. Ian Rolfe B.Sc., Ph.D., F.M.A.

Library of Congress Cataloging in Publication Data

Main entry under title:

The prehistoric world

(The Visual world)
Includes index.
SUMMARY: Describes the geological and biological evolution of the earth from its formation to the development of the human.
1. Historical geology—Juvenile literature. (1. Historical geology. 2. Earth. 3. Evolution) I. Sheehan, Angela.
QE29.P68 560 76-14898
ISBN 0-531-02448-2
ISBN 0-531-01202-6 lib. bdg.

First published in Great Britain by Sampson Low in 1975
Printed in Italy by New Interlitho, Milan

6 5 4 3 2 1

CONTENTS

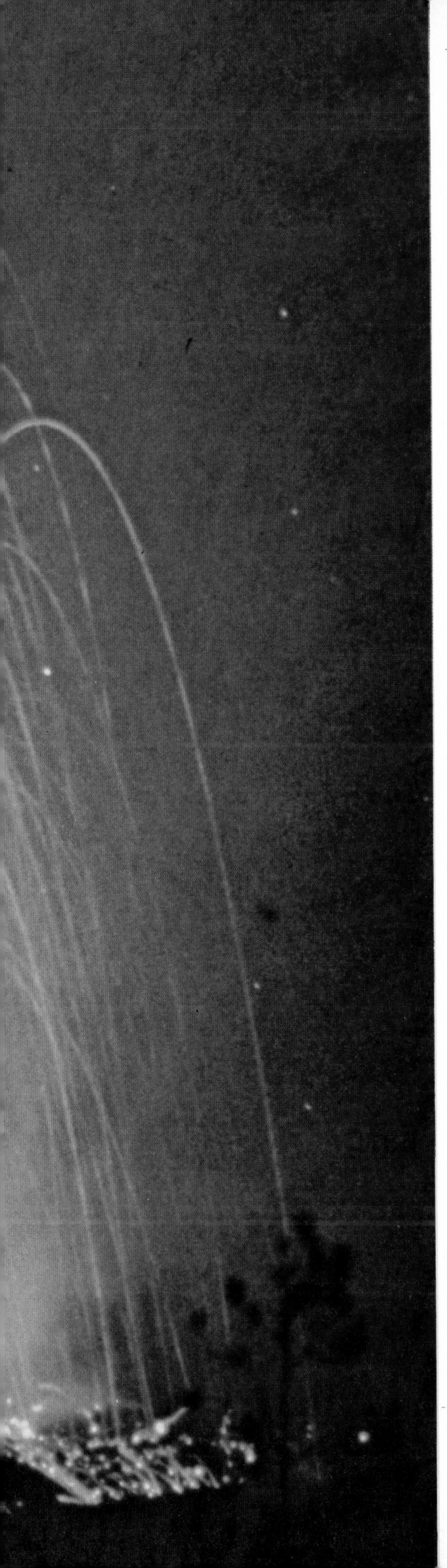

In the Beginning

Throughout most of human history, people have believed that the creation of the earth was a cosmically unique act. Yet the earth is only one of nine known planets which orbit the sun and which share so many features that there can be little doubt as to their common origin.

One of the first plausible theories purporting to explain the origin of the earth was that of the Marquis de Laplace. His 'nebular hypothesis' of 1796 maintained that the solar system began as a flattened, rotating disc of hot gases. As the disc contracted under the influence of gravity, it shed a series of concentric rings. In time, these rings condensed to form the planets, while the remaining vast cloud of gas and dust eventually became the sun. Laplace's theory explained the regular movement of the planets, but failed to show why their rotational speed was so much greater than that of the sun.

To explain the high speed of the planets in their orbits, various 'catastrophic' theories of their origin were developed. In 1916, Sir James Jeans suggested that a passing star might have drawn close to the sun and dragged a huge tongue of gas from its surface. This cigar-shaped cloud would then have broken up into fragments and condensed to form the orbiting planets. R. A. Lyttleton later suggested a variation on this theory, to the effect that the sun may originally have been a binary star – that is, that it may have had another star as its close neighbour in space. This feature is found in about one star in ten. He thought that a passing star had blasted the sun's companion from its orbit, leaving behind a mass of debris from which the planets were eventually formed. None of these theories, however,

From the culture of an ancient Indian tribe came a fanciful picture of the earth, carried by three elephants standing on the back of a tortoise. From such ideas, man has gradually progressed to a greater knowledge of the cosmos, but the origin of the earth and the life it supports is still wrapped in mystery.

succeeded in explaining why the planets orbit the sun in such a regular pattern. Furthermore, when one considers the overwhelming predominance of empty space in the universe as a whole, it is obvious that the odds are heavily weighted against a passing star approaching so close to the sun.

More recent theories have returned to Laplace's notion of a rotating gas cloud. Astronomers such as Hoyle and Weizsäcker have demonstrated that the sun's magnetic field could have set the concentric gaseous rings spinning in rapid rotation as they moved outwards from the sun. The lightest particles would have been thrown to the outer extremities of the system, forming the giant, gaseous, low-density planets (Jupiter, Saturn, Uranus and Neptune). The heaviest particles would have remained near the centre to form the small, dense inner planets (Mars, Earth, Venus and Mercury). The composition of the earth supports this theory. Compared with the sun, it is relatively poor in gases such as hydrogen and helium, but contains large quantities of heavier elements such as silicon, magnesium and iron.

The Young World

Far from being a huge sphere of solid ground, the earth is an immensely hot planet wrapped in a thin, solid crust a mere 32 to 48 kilometres (20 to 30 miles) thick. Yet our planet may not always have been the giant furnace it is today. There are two opposing theories concerning the formation of the earth. One claims that it cooled gradually from a mass of dust and hot solar gases, while the other claims that it condensed from a cloud of cold solid particles. In either case, the planet passed through an immensely hot phase at some time in its history, probably about 4600 million years ago.

If it is true that the earth was born from a cloud of cold, contracting matter, the energy which must have been released during its compression by enormous gravitational forces would gradually have melted the new planet. The infant earth thus became a molten sphere of matter, with temperatures possibly reaching as high as 4000°C. An immense additional source of heat was provided by the decay of radioactive elements such as potassium, uranium and thorium. This process is still continuing today, though only half as much radioactive material now exists.

Swirling convection currents helped to create the various layers of core, mantle and crust in the young earth. While it was still in a liquid form, the heaviest elements (iron and nickel) sank to the centre of the planet. Around this dense core there formed a compact mantle of silicon, magnesium, aluminium and other lighter elements. When the surface temperatures of the cooling earth had fallen to between 1500 and 800°C, the first floating rafts of solid crust appeared. Soon the earth was completely encased by a basalt and granite crust.

The First Seas

As the earth contracted and became increasingly dense, clouds of burning gases were sweated out of its hot interior. This searing atmosphere was a mixture of water vapour, carbon dioxide, ammonia, methane and various other gases. As the newly formed crust grew cooler, so did the atmosphere, until finally its burden of water

Forged from a swirling cloud of dust and gas in the dawn of time, the molten earth gradually cooled and condensed to a ball of solid rock. A dense cover of clouds shrouded the sweltering planet, as steam from the surface condensed. As the atmosphere grew still cooler, the clouds dropped their burden of water in a giant deluge that swamped the land and formed the first oceans.

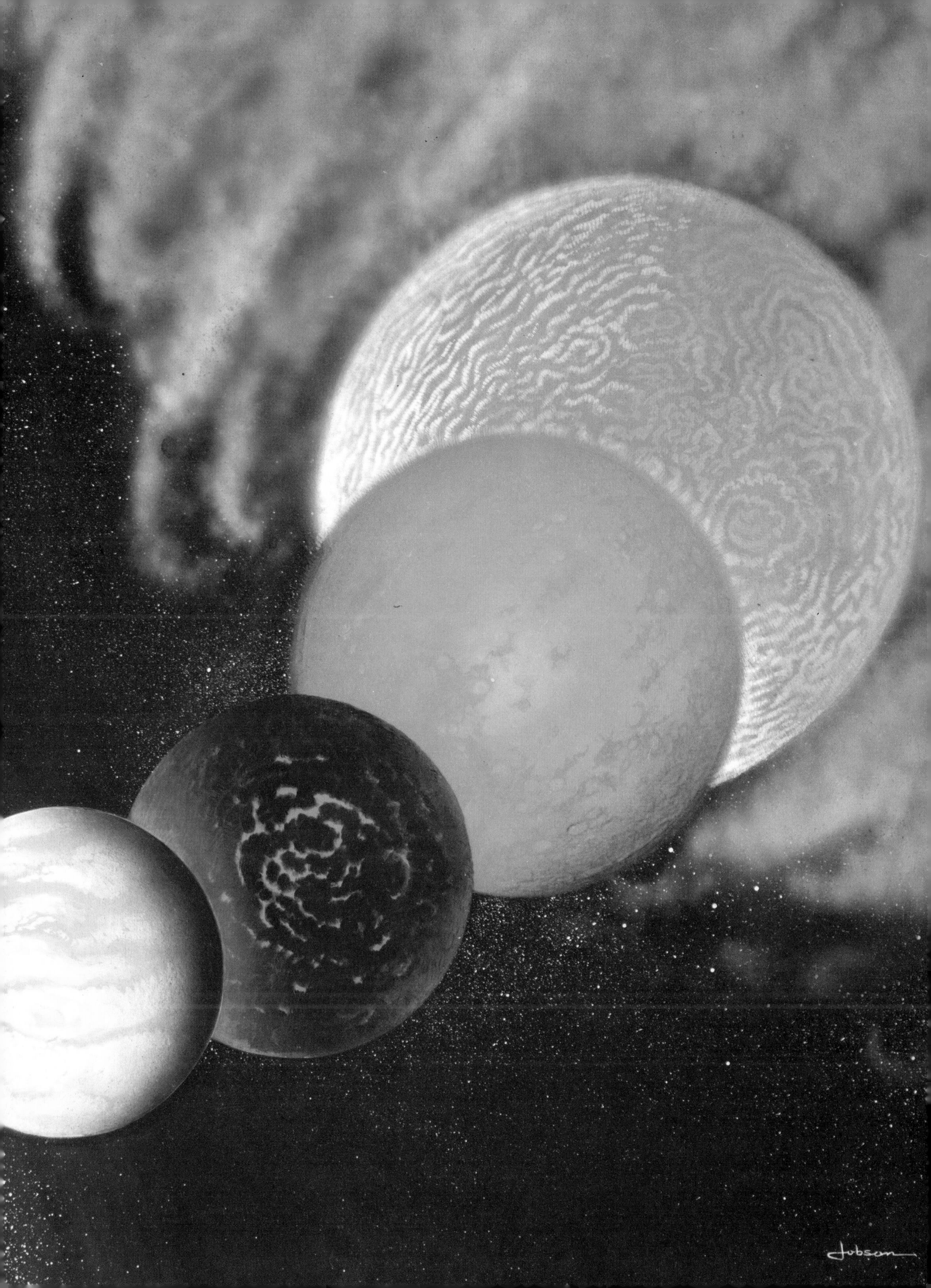

vapour could condense into rain. In a giant deluge lasting thousands of years, water poured from the skies and rushed across the steaming land in wild torrents which flooded depressions in the crust and formed the first seas. Not all the water in the modern oceans originated in this way, however. Some, which was trapped in rocks inside the crust, has been gradually released in the vapours of volcanoes.

Three Generations of Atmosphere

Hydrogen and helium, the two most common gases in the solar system, had largely escaped from the earth's field of gravity by the time its atmosphere began to cool. Thus the early atmosphere was composed mainly of water vapour, ammonia and methane. Subjected to a barrage of ultraviolet rays from the sun, the ammonia and methane were broken down into hydrogen, nitrogen and carbon atoms. Much of the hydrogen released in this way escaped into space, leaving behind an accumulating atmosphere of nitrogen and carbon dioxide. Only when the first plants capable of photosynthesis evolved did the abundant signs of free oxygen begin to appear. The plants absorbed carbon dioxide, using the carbon to construct their tissues while releasing the oxygen as waste. Gradually the earth acquired its present atmosphere in which nitrogen and oxygen form more than 99 per cent of the gases in the air.

HOW OLD IS THE EARTH?

Radiological dating has shown that the oldest rocks so far discovered on earth are 3500 million years old. Since the earth was molten for a considerable period before it cooled and formed a crust, the radiological dating of rocks cannot give a reliable figure for the age of the planet itself. However, the dating of meteorite remains is a useful alternative, since these were formed at the same time as the rest of the solar system. It has been found that their age is about 4600 million years.

Another method involves radioactive potassium 40. This decays to produce argon 40, an inert gas contained in the atmosphere. The half-life of potassium 40 is 1300 million years. By comparing the amount of argon 40 in the air with the amount of potassium 40 in the earth's crust, geologists have arrived at a figure of 4600 million years for the age of the crust itself.

Geological Clocks

Ever since Archbishop Ussher, armed only with his faith and his religious texts, attempted to calculate the age of the earth and concluded that it came into existence in 4004 BC, people have been searching for reliable ways to date the birth of our planet. One attempt involved measuring the rate at which the oceans accumulate salt. This approach led to the extremely inaccurate estimate that the seas were about 100 million years old. Another method was an attempt to estimate the rate at which a foot of sediment accumulates. This produced similarly inaccurate figures. In 1897 Lord Kelvin attempted to calculate the length of time the earth had taken to cool from a molten state to its present temperature. His figure of 20 to 40 million years was more than 100 times too low.

The efforts of William Smith (see page 18) were more rewarding. He recognized that sedimentary rocks formed during any one period of the earth's history carry a similar series of fossils. This meant that similar fossil groups found in widely separated strata could be regarded as conclusive proof that the strata belonged to the same period of geological time. With this knowledge, Smith was able to produce a comparative time scale for rocks and fossils. But he could not provide absolute dates for either, and therefore he could not estimate the age of the earth.

Opposite: In 1650, Archbishop Ussher (top) pronounced that the earth was created on 2nd October, 4004 BC, at 9 o'clock in the morning. Fortunately for him, he did not live to see the advent of radioactive dating. So far the oldest rock found in the world is a metamorphic rock from Greenland (bottom). The fact that it is metamorphic suggests that even older rocks must have existed, since metamorphism involves the transformation of other rocks by heat or pressure. So far no trace has been found of the original igneous rocks created by the cooling earth.

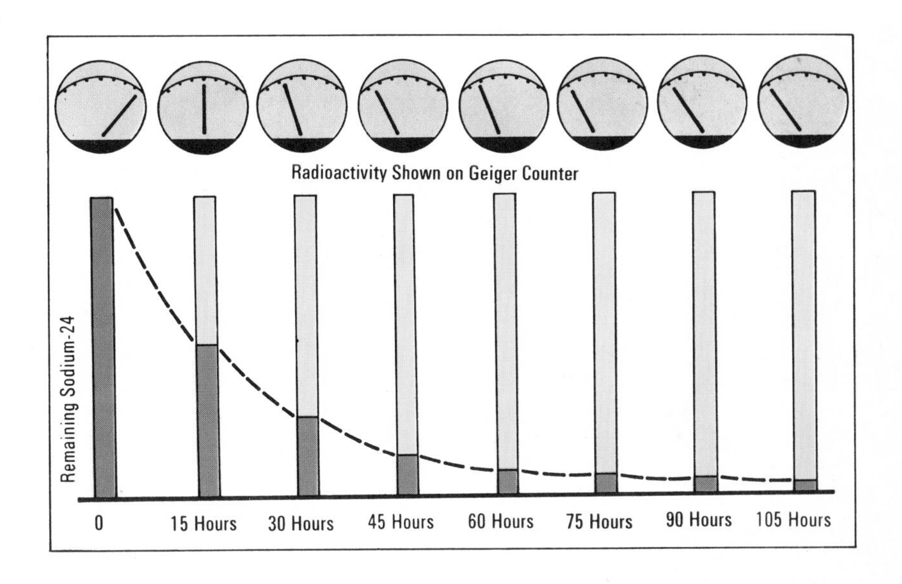

Right: The half-life period of a radioactive isotope is the time taken for half of the atoms in a sample to decay to atoms of another element. For sodium-24 this is 15 hours.

Below: A mass spectrometer, the apparatus used to measure the amounts of each element in a sample.

The formulation of an accurate absolute time scale had to wait for the discovery that radioactive elements decay at a slow and steady rate. The rate of decay is measured by the half-life of an element – the time it takes for half of a given amount of unstable radioactive material to break down into a stable form. For example, thorium (with an atomic weight of 232) decays into a form of lead with an atomic weight of 208; the half-life for this decay is 13,900 million years. Radiological dating in this instance will involve comparing the ratio of lead atoms to thorium atoms in the rock.

Even with the greatest accuracy, however, this technique always has a margin of error of 20 to 40 million years. Since the half-life of many elements is so long, this kind of dating works best for very ancient rocks: for instance, it has been estimated – allowing for the cooling of the crust – that the earth is probably some 4600 million years old.

How Fossils are Formed

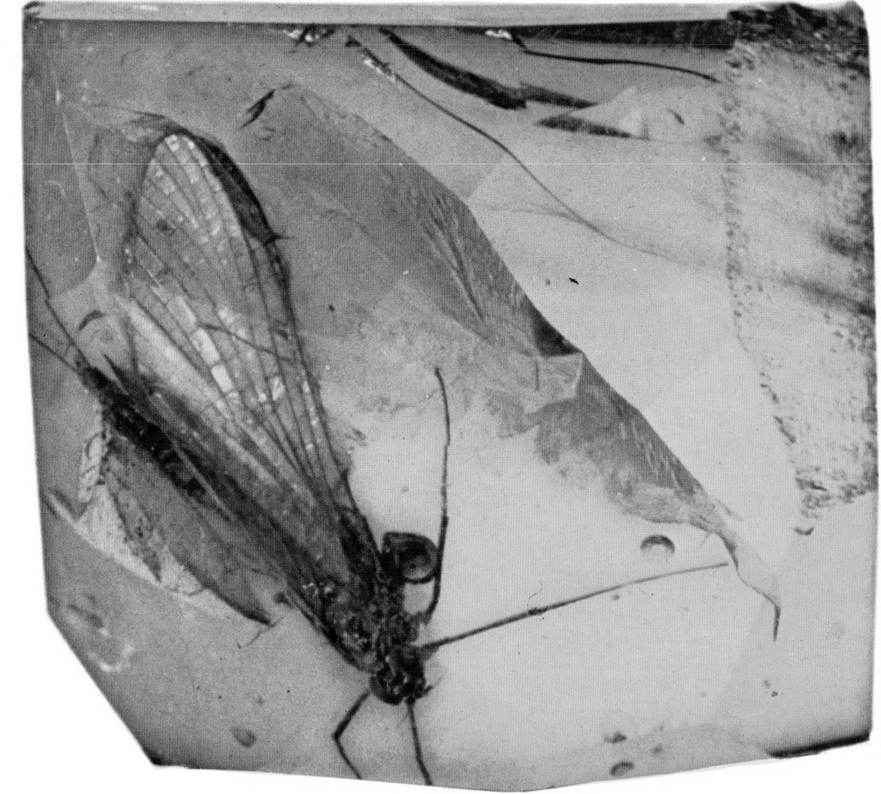

An insect almost perfectly preserved in amber for about 30 million years. Even the veins on the wings can be clearly seen.

Above: An unlikely but interesting reconstruction by Baron Cuvier of the extinct South American sloth Megatherium.

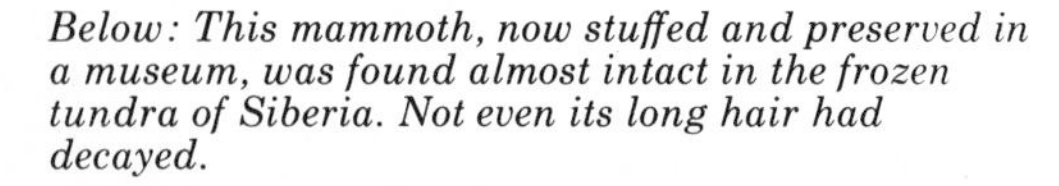

Below: This mammoth, now stuffed and preserved in a museum, was found almost intact in the frozen tundra of Siberia. Not even its long hair had decayed.

Herodotus (484–425 BC) and other scholars of ancient Greece realized that the shells which they saw in the rocks were the remains of marine creatures, and rightly deduced that the rocks had once been under the sea. But this early knowledge, together with much else that the Greeks had discovered, was eventually lost, and during the next 2000 years many strange theories were proposed to explain the existence of fossils in the rocks. One popular notion was that fossils were 'the work of the devil, placed in the rocks to deceive, mislead and perplex mankind', while an alternative theory asserted that the fossils had been created by thunderbolts. Others suggested that there was some mysterious force within the earth trying to create life, and that the fossils represented unsuccessful attempts – objects with the form of living creatures but without the vital spark of life. These strange beliefs gradually died out and people came to accept the idea that fossils really were the remains of marine creatures, but attempts to explain how fossils had found their way into the rocks produced even more fantastic theories.

Sea Mists and Floods

One suggestion put forward in the 17th century was that sea mists drifted over the land and dropped 'seeds' of sea creatures on to the rocks. These 'seeds' were supposed to have been washed down into the rocks and to have grown there. The most widespread belief, however, was that the shells and other fossils had been washed up on to the land by the biblical flood. Leonardo da Vinci (1452–1519) pointed out several valid reasons for rejecting this theory, but his writings did not reach a very wide audience and people continued to believe in the 'flood' explanation for a very long time. Not until James Hutton (see page 32) had shown convincingly that the rocks had been formed from sediments laid down on the sea bed was the true story of fossils accepted: fossils are the remains of ancient animals or plants that were incorporated into the rocks when the latter were being formed.

The Processes of Fossilization

The essential requirement for fossilization is that the plant or animal should be covered with sediment of some kind or other before scavenging animals or the natural process of decay destroy its body. This rapid burial is most likely to occur on the sea bed, where there is a constant 'rain' of debris from above, and the majority of fossils are therefore those of marine creatures. Some land-dwelling animals have, however, been washed into seas or lakes and subsequently fossilized; a few have even become fossilized after being buried in sand dunes. It is usually only the hard parts of an organism that become fossilized, because the soft parts rapidly decay or else they are crushed out of all recognition. Very occasionally, however, palaeontologists discover the fossils of soft-bodied animals such as worms and jellyfishes. These fossils are usually found in very fine-grained rocks. They consist either of very faint marks 'imprinted' by the animals when they rested on the sediments for a while or of thin films showing where the animals were squashed by the overlying sediments.

The most frequently fossilized objects are bones, teeth, and shells, together with the woody stems of plants. They can be preserved in several different ways. Teeth are very resistant to chemical change and are very often preserved in their original state, but other objects are usually altered to some extent. The shells of snails and bivalves consist mainly of calcium carbonate with a horny covering and a certain amount of organic material in the pores. When one of these

ROBERT HOOKE

These drawings of ammonites and corals were made in the 17th century by Robert Hooke, an outstanding physicist and mathematician. Like so many scientists of his time, he was not content with just one or two branches of science. On several occasions he turned his attention to geology and palaeontology and used a microscope to investigate the detailed structure of various fossils. Examining ammonite fossils, he realized that the shells had contained several chambers and that the animals were related to today's pearly nautilus.

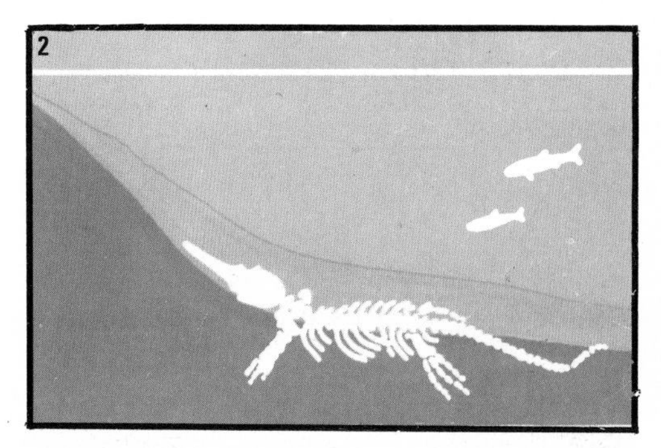

The history of a fossil. A marine reptile dies and its body sinks to the sea bed (1). Its flesh decays, but the skeleton remains (2). The bones are gradually petrified and the sediments harden into rocks. Much later, earth movements cause the sea bed to buckle. The rocks are lifted up out of the sea (3) and the processes of erosion begin. Layer after layer is stripped off and the fossilized remains of the reptile are eventually exposed (4).

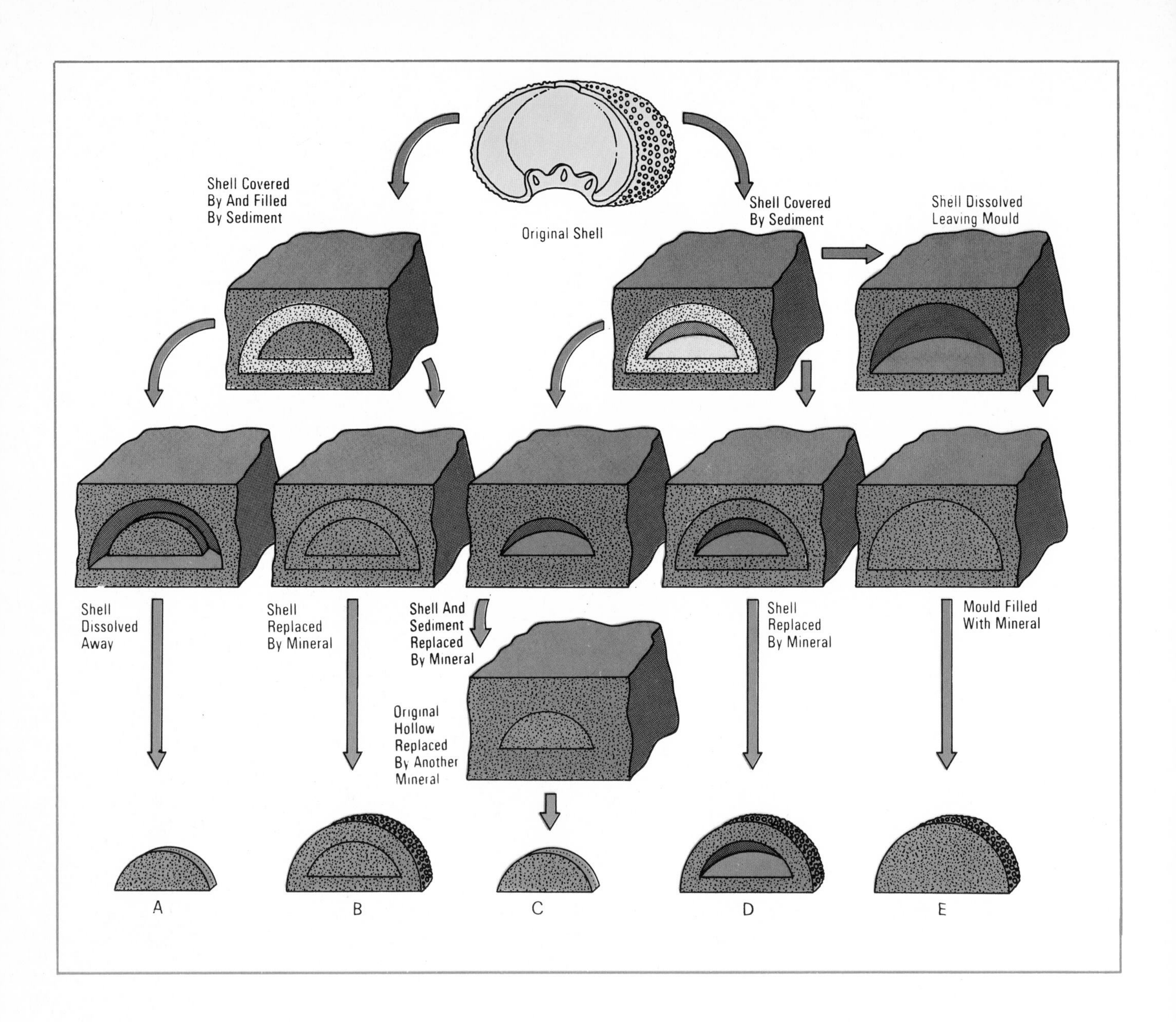

A diagram showing some of the ways in which a hollow sea urchin shell can leave fossils in the rocks. Some of the fossils show the form of the outer surface of the shell, and some show what the inner surface was like. Fossil A is a core, or filling, whose outer surface bears the pattern of the original inner surface of the shell. Fossil B is a petrified shell or cast with a filling of sediment (rock). The internal and external features of the shell are preserved. Fossil C is a core, just like fossil A. Fossil D is a petrified shell, or cast, just like fossil B except that it is hollow. Fossil E is a solid replica of the sea urchin. It bears the pattern of the original outer surface of the shell, but no other features are preserved. The fossil does not tell us how thick the original shell was.

These five kinds of fossils all leave hollow moulds when they are released from the rocks, but the moulds are not all alike. The moulds left by fossils A, B, D, and E are all of the outside surface of the shell, but the mould left by fossil C bears the pattern of the inside of the shell.

A solid object cannot be fossilized in so many ways because there is no hollow to form a core, or filling.

shells is buried, the organic material decays or is dissolved away. Relatively 'young' fossil shells are often found in this state. In older fossils the pores have been filled with minerals gradually deposited by water seeping through the rocks. These minerals include silica, iron pyrites, and calcite – a very hard form of calcium carbonate. This process is called *permineralization.* It generally makes the shells much harder and more resistant to destruction. The same process often takes place in fossilized bones.

The original material of a shell or a bone is very often completely removed and replaced by mineral matter during the fossilization process. The object is then said to have been *petrified* or turned to stone. Some of these petrified fossils still show their internal structure, because the original material was replaced molecule by molecule. Among the best known of these fossils are the petrified tree trunks and 'forests' which exist in various parts of the world. It is possible to identify the conducting tubes and annual rings in these fossilized trees, just as it is in freshly felled tree trunks. The majority of petrified fossils, however, reveal no internal structure, although they may portray the surface details very accurately.

The removal of the original bone or shell material by percolating water may not be followed immediately by the deposition of new minerals, and in such instances there will be a hollow in the rock. Such a hollow is known as a *mould* and, although it is not actually a part of the original animal, it is a true fossil because it tells us a great deal about the animal. If, at a later date, the mould becomes filled with another mineral, the mineral will take on the shape of the original object and it will qualify as a petrified fossil. Fossils formed in this way are called *natural casts*. If the rock containing the cast is broken open, it may be possible to remove the cast from its mould and to get two fossils for the price of one. By pouring plaster or some other material into the mould, it is possible to make artificial casts of the fossil. Many of the fossils seen in museums are actually artificial casts made in this way, for not every museum is able to acquire genuine specimens of every kind of fossil.

Two fossils of the Cambrian trilobite Agnostus: a cast (left) and a mould (right).

Carbon Copies

Plant remains and other objects consisting largely of organic (carbon-based) material frequently undergo a process called *distillation*, brought about by the immense pressures of the overlying sediments. These pressures drive off much of the oxygen, hydrogen, and nitrogen and reduce the original object to a residue consisting mainly of carbon. Plant leaves and the horny skeletons of graptolites (see page 62) have usually been fossilized in this way, and their residues generally form delicate films on the rock layers. Our coal seams have been formed in the same way from the vast accumulations of tree trunks and other plant material which sank into the swamps of the Carboniferous Period. The best coal is that which has been most deeply buried or which has suffered the greatest pressures through earth movements. This coal has lost nearly all of its gaseous materials and consists almost entirely of carbon. It therefore gives out a great deal of heat when it is burned. Coal seams are found in many parts of the world, including Antarctica. These Antarctic coal deposits indicate that the continent once possessed a much warmer climate and a very luxuriant vegetation.

The fossilized leaf of a Carboniferous plant, preserved as a film of carbon.

Fossils in Tar, Ice, and Amber

The majority of fossils are found in ordinary sedimentary rocks which have been formed by the accumulation of layer upon layer of fine rock debris. There are, however, a number of other situations in which plants and animals have been preserved and fossilized.

In a part of Los Angeles known as Rancho la Brea (see page 146) there are a number of small lakes, but these are no ordinary lakes: they are pools of tar, and they have yielded a remarkable array of fossilized bones. The tar has been seeping out of the ground for thousands of years and the pools are quite deep, although much of the tar has now solidified. Rainwater accumulates on top of the tar and forms shimmering lakes. These lakes are now fenced to prevent accidents, but in prehistoric times there were no fences and many animals came to drink there. It was the last drink for each of them, for they soon became trapped in the sticky tar. The struggles of the doomed animals obviously attracted carnivores and scavengers, and these also became trapped in their eagerness to get at the feast. A vast assemblage of animals thus became entombed in the tar pits. Their soft parts soon disintegrated, but their bones were perfectly preserved, giving palaeontologists a wonderful chance to learn about the animals that roamed North America during the last 150,000

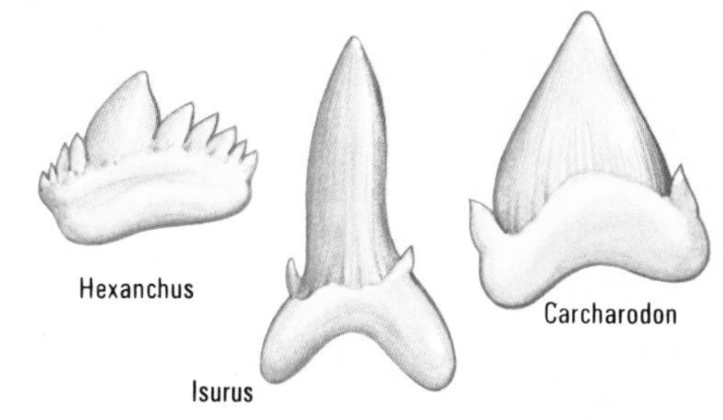

Three shark teeth from Cretaceous rocks. These very hard and resistant objects have been altered very little during the 100 million years of preservation.

These ammonite fossils from the Jurassic clays are cores, or fillings, made of iron pyrites (fool's gold).

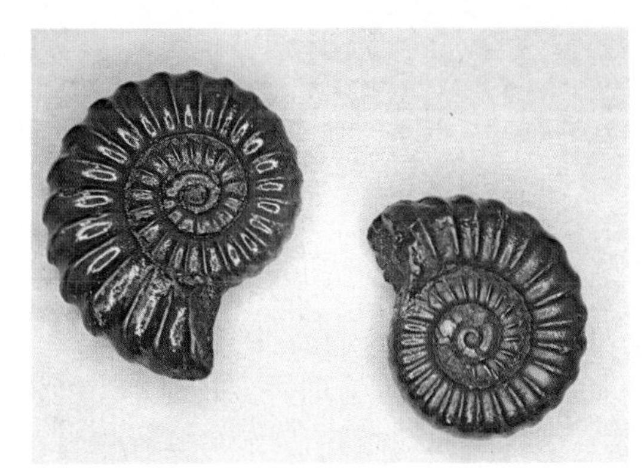

THE FATHER OF ENGLISH GEOLOGY

Apart from their inherent interest, fossils are very important in helping geologists to unravel the history of the earth. One of the first people to realize the importance of fossils in this field was William Smith. He was born in Oxfordshire in 1769 and among his first toys were the bun-shaped 'pound stones' which his mother used as weights in her bakery. These were really the fossils of sea urchins from the local rocks and William soon began to collect them himself.

Smith became interested in all kinds of fossils and he continued to collect them when his work as a surveyor and engineer involved him with coal mines and with the excavations for Britain's expanding network of canals. He was very methodical in his collection, and took care not to mix up fossils from different rock layers. It then became obvious to him that each rock layer or formation contains its own particular assemblage of fossils, and, therefore, it should be possible to identify a rock formation wherever it occurs by looking at its fossils. This discovery became the basis of the geological time scale (see page 36) and of all geological mapping.

In 1815 William Smith himself produced the first large-scale geological map by examining fossils in quarries and cuttings all over the country. The fossils told him which rocks occurred at each place, and he was then able to draw up his map. He also made very careful drawings of the fossils associated with each rock formation and published them in 1816 in a book called *Strata Identified by Organised Fossils*. The maps which Smith produced were of great interest to geologists and naturalists studying fossils, but their value extended far beyond that. By studying the maps, Smith himself could work out the best route for a new canal – always best sited on impervious clay rocks which hold up the water. The maps also revealed the most likely sources of building stones in an area, and they were of immense help to engineers concerned with mining or with water supplies.

The Geological Society of London awarded William Smith a gold medal in 1831 and, in recognition of his work, the President of the Society called him 'The Father of English Geology'.

A petrified tree trunk in which the thickness of the bark and some of the annual rings can still be made out, although the original material has been completely replaced by minerals.

years. Among the animals whose skeletons have been found in the tar pits are mammoths, mastodons, wolves, foxes, tapirs, and sabre-toothed tigers. There are also many skeletons of vultures and other birds. Similar tar pits have been discovered in Peru, and these have yielded the remains of many strange South American mammals.

Today we use refrigerators to prevent meat and other foods from decaying, but nature was preserving things by refrigeration long before men ever thought of doing so. Several mammoths have been found embedded in ice and frozen soil in the Siberian tundra. Although they must have been there for more than 30,000 years, their coats were still in good condition.

Amber, the hardened resin of pine trees and other conifers, is usually found as smooth, yellowish pebbles on the beach. Much of it is between 10 million and 50 million years old. Insects, spiders, and other small animals are occasionally seen inside the amber pebbles, almost perfectly preserved or fossilized. These small animals were obviously attracted to the original resin when it was oozing from the tree trunks, and became trapped in it just as animals do today.

Peat, which is composed of the partially decayed remains of plants, is also a good preservative and many ancient bones have been discovered in peat bogs. Pollen grains are also preserved in the peat and, by studying and identifying the pollen found at various levels in the peat bogs, scientists can learn a great deal about the vegetation of the past. Existing peat bogs are generally rather young in geological terms, dating back only to the end of the last ice age, and many geologists consider that the animal remains in them are not yet old enough to be true fossils. They are therefore called sub-fossils, but they are certainly fossils in the making and they will be true fossils one day. The peat itself will also become fossilized through the process of distillation and, if sufficient pressure is applied, it will eventually become coal.

Rocks Made of Fossils

Fossils are not evenly distributed through the rocks because the conditions necessary for fossilization have not always existed. Some rock layers are completely devoid of fossils, whereas others are so crowded with them that they are literally made of fossils. Some of the limestones laid down during the Carboniferous and Jurassic Periods are of this type. They were formed at times when the land was rather low-lying and was sending relatively little sediment into the sea, so that the sea-bed deposits consisted mainly of the shells and skeletons of animals together with variable amounts of calcium carbonate powder. The latter was derived mainly from crushed shells and from lime-secreting algae, but some must also have been deposited by the warm, shallow water, just as lime is deposited in kettles and water pipes. The powder gradually crystallized into the very hard form of calcium carbonate called calcite, and it cemented the shells and other fragments into rocks. Many of the Carboniferous limestones contain crinoid, or sea-lily, skeletons in large quantities (see page 64), while others contain little but brachiopod shells (see page 67). At certain levels, the rocks are composed almost entirely of coral skeletons fossilized in large reefs just where they grew (see page 68). Some of the Jurassic limestones also consist mainly of fossilized coral reefs.

Dense beds of shells or bones can be found in a variety of rocks. Colonies of bivalves living on the sea bed have often been overwhelmed and killed by sudden deposits of sand or mud brought down in times of flood. These colonies then became fossilized where they grew, and the fossilized shells can be found nestling in slabs of rock. Fossilized groups of this kind can be recognized quite easily, because the shells are usually of several different sizes – corresponding to the various ages of the animals in the colony – and most of them are complete. The majority of shelly beds, however, have been laid down

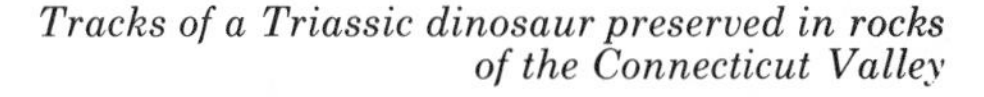

Tracks of a Triassic dinosaur preserved in rocks of the Connecticut Valley

MARY ANNING THE FOSSIL HUNTER

Visitors to Lyme Regis and Whitby in England are often surprised to see large circular stones offered for sale in the shops. These stones are ammonite fossils from the Blue Lias rocks which form the cliffs in these seaside towns. The trade was first started by a carpenter called Anning who lived in Lyme Regis at the beginning of the 19th century. He had a flair for finding fossils in the cliffs and rocks along the seashore, and this flair was inherited by his daughter Mary who accompanied him on many of his collecting trips. Mary was only ten years old when her father died, but she continued to collect fossils and sell them to support her family. Among the fossils she found, there would undoubtedly have been many petrified ammonites and sea urchins from the limestone bands in the rocks, and she would also have picked up numerous pyrite fillings of ammonites (see page 17). But it was her discoveries of fossil reptiles that really made her famous.

When she was only eleven years old Mary found the first more or less complete fossil of an ichthyosaur (see page 102), and she hired a team of men to help her extract the bones from the rocks. A few years later Mary Anning became the first person to discover a complete plesiosaur fossil (see page 102), and she was also the first person to find a pterodactyl in Great Britain. Most of her discoveries were made in the Blue Lias rocks around her home town, and it is said that one of her greatest discoveries was made when she climbed on to a boulder and then looked down to see that she was standing on the fossilized skull of a reptile.

Although not a scientist, Mary Anning taught herself a great deal about animals so that she could understand the fossils which she was so good at finding. Many famous geologists came to visit her to learn the art of fossil collecting.

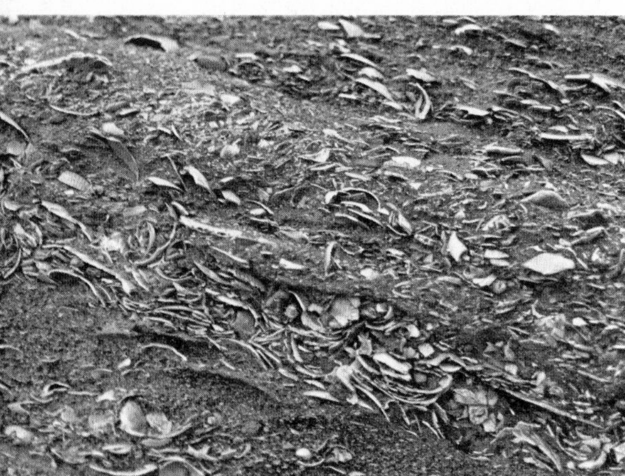

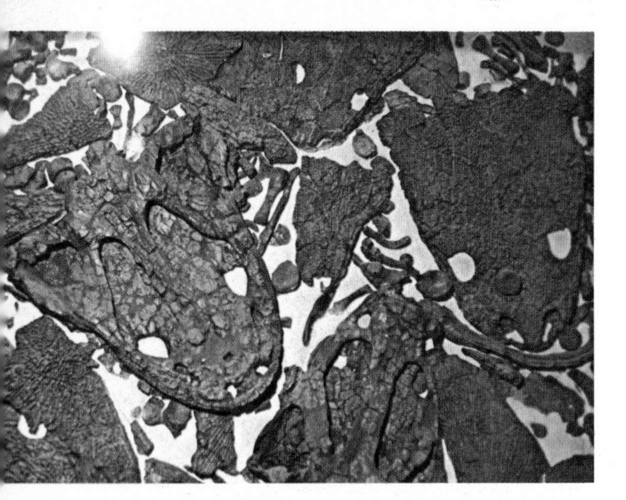

Above: A dense bank of shells and shell fragments on a modern sea shore (top). If the shells become covered with sand they may become fossilized and converted into rocks like the Red Crag (centre). This rock, laid down in the early Pleistocene Epoch, consists of layers of shells separated by beds of sand. The particles are only loosely cemented together and the rock crumbles very easily. Crags form low cliffs on many stretches of the East Anglian coast in England and they are gradually crumbling into the sea again in several places. Occasionally, beds of rock are found crammed with the fossilized bones of much bigger animals, such as these amphibians (bottom). The animals probably died when their Triassic pond dried up.

Digging up a dinosaur in Mongolia (opposite) involves much patient work. The rock is chipped and brushed away piece by piece until all the bones are exposed just where they were found. In this way, palaeontologists can learn just how the bones were placed. Photographs are then taken on site before the bones are removed and taken to the laboratory. There, plaster casts can be made (below) and these help in reconstructing the original appearance of the animal.

as beach or sand bank deposits and most of the shells are broken or worn, although they do tend to be of more or less the same size in any given area. You can often see these shelly beds in the making if you go to the sea shore: the waves tend to concentrate shells into distinct bands along the shore, and it is easy to see how these bands can then be covered with sand and fossilized. This has happened many times in the past, especially during the formation of the crag rocks on the east coast of England. Some of the crumbling cliffs in this region are more than 12 metres (40 feet) high and they consist entirely of thin beds of shells separated by narrow layers of sand.

The bone beds which occur in certain rocks were formed in the same way as the shell beds – through the concentration of assorted and often very worn bones by the waves. Very occasionally, however, palaeontologists come across layers of complete skeletons where a large number of animals have been fossilized more or less on the spot. Such fossil groups usually result from the death of numerous fishes or other animals in drying lakes.

Where to Find Fossils

Although only a very small proportion of plants and animals ever become fossilized, the rocks do contain immense numbers of fossils which are quite easy to find if you know where to look. It is obviously no good looking for fossils in granites and other igneous rocks which have been formed from molten material deep in the earth, and it would almost certainly be a waste of time to hunt for fossils

in the lavas that have been thrown out by volcanoes. There are rare instances where animals have been preserved after being overwhelmed by volcanic ash and relatively cool lava, but most living creatures meeting their end in this way would almost certainly be completely destroyed. Natural moulds in the volcanic ash preserve the shapes of the unfortunate inhabitants of Pompeii who were swallowed up in the ash which rained down from Vesuvius in AD 79, but these are not yet old enough to be true fossils.

The rocks in which to search are sedimentary rocks, which have been formed from particles laid down long ago on the sea bed or elsewhere. The finer-grained rocks – the clays and shales, and the limestones – are the most rewarding. Fine sandstones may also yield fossils, but the coarser ones are unlikely to be worth examining. They are so porous that any fossils that may have been present have usually been dissolved away by the water, and the coarse particles do not preserve very faithful images of the original shells. A geological map will help you decide where to search.

The best places in which to look are cliffs, road and railway cuttings, roadworks, and even building sites – anywhere, in fact, where the rocks are exposed below the soil. Permineralized and petrified fossils are often much harder than the surrounding rocks and they often stand out from cliff and quarry faces when the rock particles have been weathered away. The fossils eventually fall out and rich pickings can often be found at the bottom of the cliffs. Shiny ammonite fillings made of iron pyrites are very commonly

Below: A palaeontologist works painstakingly on a plaster cast of the inside of the skull of a fossil man. Such casts reveal a great deal about the size and structure of the brain, and from them experts can deduce something about the intelligence and habits of our early ancestors.

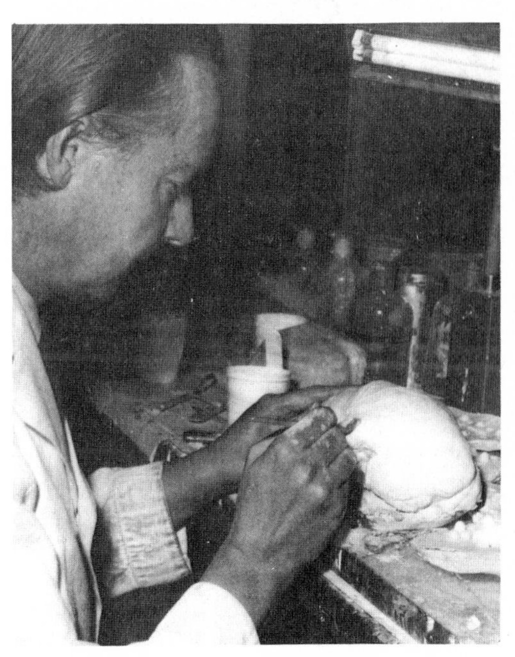

1

2

3

The dinosaur skeletons which we can see in the world's museums have often been dug up in remote regions and enormous efforts have been put into recovering them. The giant Brachiosaurus skeleton (opposite), towering above the other fossils in the Berlin Museum, was discovered at Tendagaru in East Africa in the early years of this century. Many other dinosaurs were found there and a German expedition spent about four years digging some of them out. Pictures from the site show some of the work involved:

1. *The foreman of the expedition exposes a huge dinosaur rib.*
2. *Werner Janensch, the leader of the German expedition, discusses a new find with an assistant.*
3. *The huge mounds show just how much earth had to be removed to expose the dinosaur bones.*
4. *The dinosaur bones, securely encased in plaster, begin their long journey by hand, ship and rail to Berlin for re-assembly.*

The small drawing inset (opposite) was made by one of the native workmen at Tendagaru to show how he thought the whole skeleton might look.

4

You can find fossils in the most unlikely places. High up on the cathedral of Notre Dame in Paris, this limestone gargoyle is subjected to severe weathering. Small fossils, softer than the surrounding stone, have dissolved away and left numerous holes (moulds) in the surface.

washed out of Jurassic clays, while flint fillings and replicas of sea urchins are abundant in chalk cliffs. Large flints and other boulders are always worth attacking with a geological hammer, for there is very often a fossil nestling inside. Where possible, rocks should always be attacked along the bedding planes. These are the dividing lines between the various layers of sediment and it is in these planes that the original shells or other objects would have settled.

Chipping Them Out

Many fossils can be collected with no equipment at all, but a good hammer is necessary if you are going to collect fossils seriously. It should be one made specially for geologists, with a handle strong enough to withstand repeated blows on the hardest rock. The head should also have one chisel-shaped end, which allows you to prise off loose flakes of rock. An all-steel chisel or bolster is also a useful tool for chipping fossils out of the rock. Some of the fossils come out very easily, especially from the softer clays and shales, but others are firmly cemented to the rock and you will have to take a block of the rock as well. Careful treatment with acid may dissolve away unwanted rock, but this is really an expert's job. As long as the fossil is clearly visible, a piece of rock around it is not a bad thing – it protects the fossil and also tells you immediately what kind of rock it came from. The fossil can be cleaned up with an old toothbrush and a toothpick, and you must then label it with the locality from which it came. Fossil shells are often very fragile, and they may be cracked when you find them. A coat of varnish sprayed on to them may harden them, or you can take an impression with modelling wax before trying to extract them. The impression can later be used to make an artificial cast with plaster. If you are lucky enough to find a really large fossil, such as a complete fish or an ichthyosaur – or even a dinosaur skeleton – do not try to dig it out yourself. Contact your local museum and tell them about the fossil. They may be able to arrange for experts to extract it in the way shown on page 21.

FOSSILS TRUE AND FALSE

Dendritic (shrub-like) pyrolusite crystals branching out on a piece of flint and looking like fossilized plants.

The word 'fossil' literally means 'dug up'. It was originally used to describe anything unearthed from the ground. Today, however, the term is applied only to the remains or traces of organisms that lived in earlier epochs of geological time. The essential thing is that the fossil must give some indication, however small, of the size or form of the organism. Footprints left in ancient muds which later hardened into rocks are thus true fossils, and so are the tunnels of various worm-like creatures. Even dung can be fossilized if it becomes impregnated by minerals. Fossilized droppings are called *coprolites*; they often reveal what ancient animals ate.

Rocks also contain many other markings and objects which are not fossils, although they may reveal information about past conditions. The cracks in sun-baked mud, the ripples in the sand of an ancient seashore, and the pits made by rain drops can all be preserved when sediments harden into rocks. Among the most fascinating of geological finds are the various mineral concretions which occur here and there. Flints, for example, take on some very weird shapes and it is easy to mistake them for fossils. Certain minerals, notably the manganese compound known as pyrolusite, crystallize in branching threads which spread out through cracks in the rocks. Inexperienced fossil hunters can easily mistake these branching minerals for the carbon films of fossilized plants.

Charles Darwin (1809–1882)

Evolution

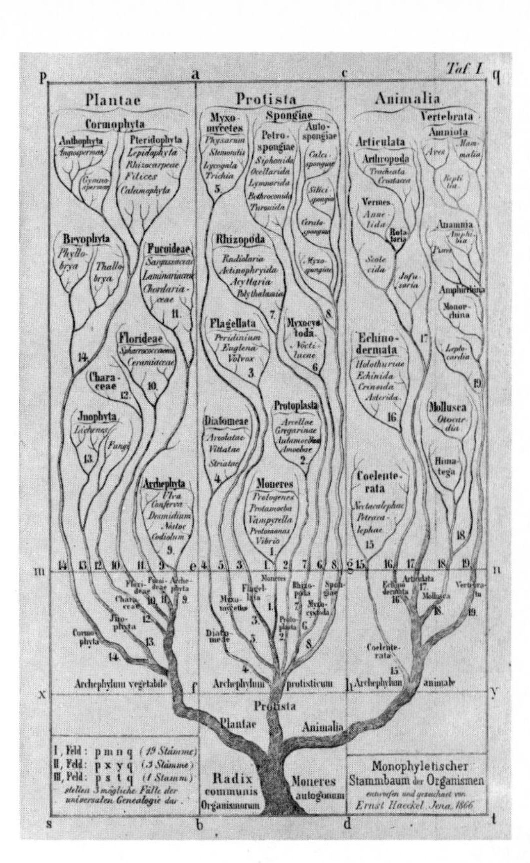

Part of a 'family tree' drawn in 1866 by the German biologist Ernst Haeckel to show how various animal groups were related by evolution. Compare this with the chart on pages 30–31.

Alfred Russel Wallace (1823–1913)

The fossils preserved in the rocks show very clearly that the plants and animals of the past were not the same as those living today. They also show us that the organisms living 100 million years ago were very different from those living 500 million years ago. The explanation is that the plants and animals have evolved: the ancient organisms varied slightly from generation to generation and, over millions of years, one kind or species gradually changed into a new one. This is what we mean by evolution, and it is still going on.

Although the theory of evolution is accepted by almost everyone today, people did not really begin to believe in it until the middle of the 19th century. Before that it was generally thought that our plant and animal species had been created in their present forms and that they could not change. This theory of special creation was accepted by many famous scientists, including Georges Cuvier of France. In order to explain the existence of different fossils in each rock layer, Cuvier invented the *catastrophic theory*. He suggested that floods and earthquakes periodically devastated parts of the earth and destroyed living things in those areas, and that new forms of life from other parts of the world later repopulated them.

Jean Baptiste Lamarck (1744–1829)

The Evidence for Evolution

The theory of special creation was not supported by any real evidence, but there is a wealth of evidence supporting the idea of evolution. Predictably, some of the strongest evidence comes from the fossils themselves. For example, palaeontologists have unearthed long sequences of sea urchin fossils ranging through many rock layers and representing many millions of

Georges Leopold Cuvier (1769–1832)

SELF IMPROVEMENT

Jean Baptiste Lamarck was a French naturalist and a fervent believer in evolution. He was, in fact, one of the first to suggest a way in which evolution might work, but his ideas on this subject were, and still are, unacceptable to science. He suggested that individual animals had an internal urge to improve themselves and that they would develop new features *in order to* cope with new conditions. He also suggested that, having acquired new features during their life-times, the animals would pass them on to their offspring. To give a simple example, Lamarck would have said that the black peppered moths (see page 29) turned black during their lives because it was helpful to be black. In fact, of course, the black moths appeared quite naturally by mutation (see page 29) and the black colour just happened to be useful. If the black form had appeared 100 years earlier, before pollution began to blacken the trees, the mutation would have been positively harmful. There is no evidence to support either the 'internal urge' theory suggested by Lamarck or the inheritance of features acquired during a life-time, and Lamarck's theories have been discarded. Lamarck did, however, make some significant contributions to the study of fossil invertebrates and wrote some important books on the subject early in the 19th century.

Four Toes

EOCENE

Hyracotherium (left), the four-toed, terrier-sized ancestor of the horse, was a forest browser. *Orohippus*, from the middle, and *Epihippus*, from the late Eocene, were slightly larger descendants.

Three Toes

OLIGOCENE

Mesohippus (left) had only three toes. Its slightly larger descendant *Miohippus* was still a browsing animal but it could run fairly well.

THE HISTORY OF THE HORSE

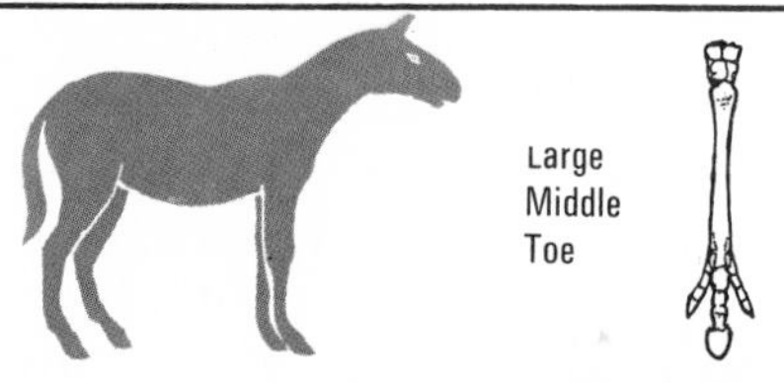

Large Middle Toe

MIOCENE

Towards the end of the period, a larger, more efficient runner called *Merychippus* emerged. With side toes considerably reduced, it walked on its hoofed middle toe. It was a grass-eater.

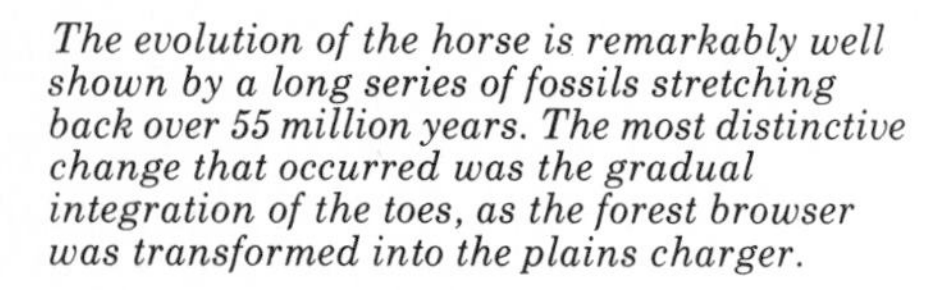

The evolution of the horse is remarkably well shown by a long series of fossils stretching back over 55 million years. The most distinctive change that occurred was the gradual integration of the toes, as the forest browser was transformed into the plains charger.

One Toe

PLIOCENE

Two new single-hoofed forms arose from *Merychippus*. These were *Pliohippus* (left) and *Hipparion*, which spread from North America to the Old World.

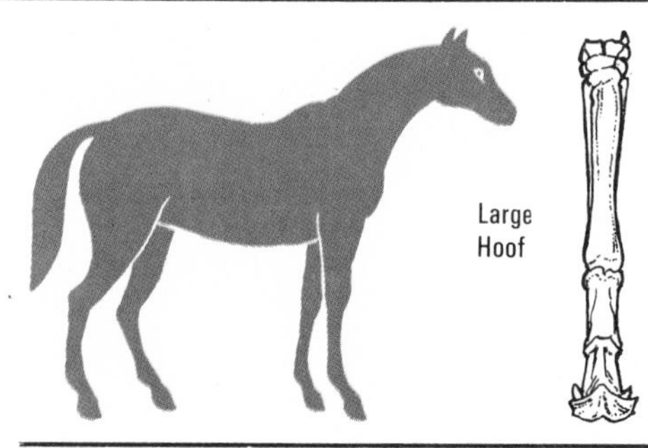

Large Hoof

PLEISTOCENE

From *Pliohippus*, came the large, fast-moving modern horse, *Equus*, and *Hippidium*, which evolved in South America but became extinct before the end of the epoch.

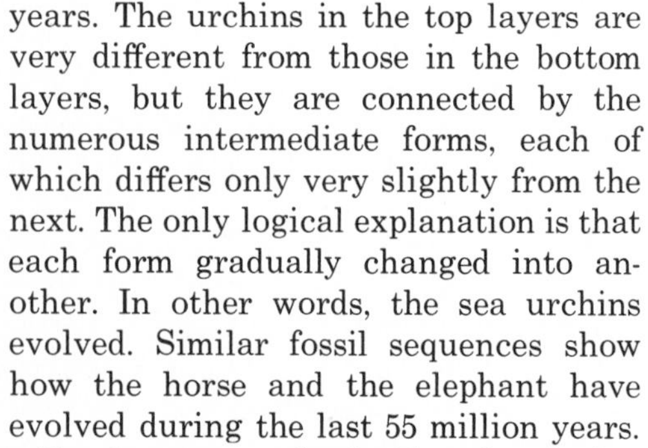

years. The urchins in the top layers are very different from those in the bottom layers, but they are connected by the numerous intermediate forms, each of which differs only very slightly from the next. The only logical explanation is that each form gradually changed into another. In other words, the sea urchins evolved. Similar fossil sequences show how the horse and the elephant have evolved during the last 55 million years.

Further support for the idea of evolution comes from the structural similarities between different animals. A look at the front limb-bones of a dog and a bat will show that both are built on the same plan, although the limbs are used in totally different ways. If these animals had been specially created, why should they have been given the same skeletal structure? The only logical explanation of the similarity is that the two kinds of animals have descended, with modifications, from a common ancestor.

Equally strong evidence is provided by the embryos of backboned animals. While developing inside their eggs or inside their mothers, birds and mammals pass through stages which are almost identical to those observable in reptiles and amphibians. This suggests very strongly that the animals have all descended from a common stock and that they have not been created independently.

CONVERGENT EVOLUTION

Because the environment controls the direction of evolution, it follows that animals living in similar places and leading similar lives will develop similar features. This is true even if the animals live in different parts of the world and belong to different groups. This phenomenon is called *convergent evolution*. Some of the best examples involve the cacti of the American deserts and the spurges of southern Africa. Although quite unrelated, these plants show marked similarities because both groups are adapted for life in the desert. South America was isolated for much of the last 65 million years and many strange mammals evolved there. Some of them, however, came to resemble mammals living elsewhere because they occupied similar habitats. *Thoatherium* was a small pony-like mammal living in South America in Miocene times. It was not related to the horses, but it had similar habits and its legs (right) became remarkably similar to those of the North American horses.

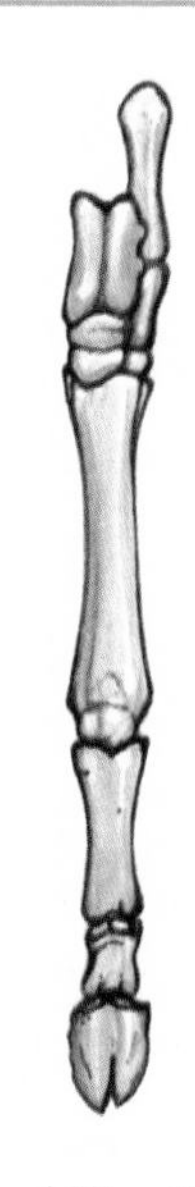

The leg of Thoatherium.

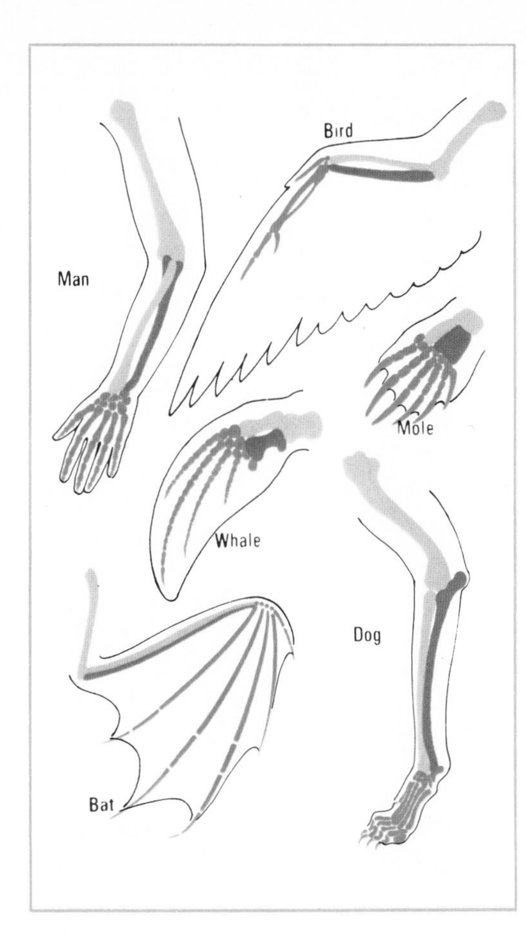

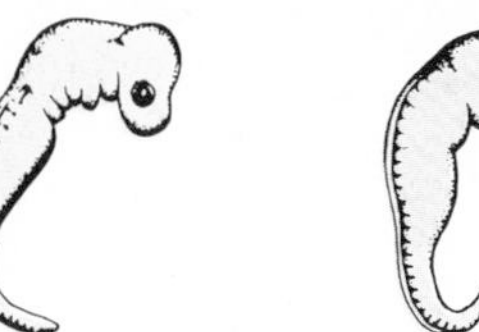

Above: At an early stage, vertebrate embryos are hardly distinguishable. From left to right are: a fish, a reptile, a bird and a mammal.

Left: The front limbs of various vertebrate animals showing how they are all based on the same bone structure, despite differing adaptations for flying, walking or swimming. This is strong evidence that they have all evolved from a common ancestor.

The geographical distribution of plants and animals also suggests that evolution has taken place. The Hawaiian Islands, for example, support several kinds of birds called honeycreepers. The birds differ in their beaks and feeding habits, but in other respects are rather similar. This suggests that they evolved from a common ancestor that managed to reach the Hawaiian Islands in the past and then evolved into the different kinds of honeycreeper. This is an example of *adaptive radiation* (see page 31), with all the birds spreading out and becoming adapted to all kinds of available food.

A final example in support of the theory of evolution is the existence of numerous cultivated or domesticated varieties of plants and animals. These varieties have been produced by human intervention, but they show that plant and animal species can, and do, change or evolve.

How Evolution Works

Charles Darwin will always be regarded as one of the most famous biologists in history, because it was he who first explained how evolution works. It is important to realize, however, that Darwin was not the first to suggest that evolution had taken place. Many earlier thinkers, including some of the ancient Greeks, had put forward the idea. They realized that animal and plant life had become more and more complex during the earth's history, but they were unable to give a satisfactory explanation of the continued progression. Why didn't some of the organisms evolve the other way? Darwin provided the answer in his theory of *natural selection*, and the way was then open for general acceptance of the idea of evolution. In 1831, when he was only 22 years old, Darwin joined HMS *Beagle* as a naturalist and began a round-the-world voyage which was to last almost five years. During the voyage he studied the distribution of various plant and animal groups, and noticed that island animals were similar to, but not exactly like those of the nearest mainland. In South America, he found fossils which were obviously related to living forms although they differed from these forms in some respects. He also had ample opportunity to see how

Below: Four species of Hawaiian honeycreeper. Their beaks have evolved in different ways in response to different diets.

A giant tortoise from the Galapagos Islands. Giant tortoises are found on several remote islands, for they survive a long sea voyage better than smaller tortoises. Once arrived, they also find few natural enemies.

WHAT IS A SPECIES?

A species is an individual kind of plant or animal. The members of a species are all able to breed with one another and produce fertile offspring. They all look more or less alike as a rule, but there is some variation and some varieties of a species are amazingly different. Cauliflowers do not look much like cabbages, for example, but they can breed together and so they are regarded as varieties of a single species. It is more difficult to decide whether two fossils belong to the same species or not because it is not known whether they could breed with each other.

admirably plants and animals were adapted to their own environments and their ways of life. He became firmly convinced that evolution was, indeed, the explanation. After his return to England, Darwin continued to collect more evidence to support the idea of evolution, and he was eventually able to explain exactly how it had taken place. His theory of natural selection can account for evolution in both plants and animals. It can account for the development of the marvellous examples of camouflage which we find among animals, and also for the evolution of new species and groups.

The theory of natural selection can be summed up in two familiar expressions: the *struggle for existence*, and the *survival of the fittest.* Plants and animals always produce far more offspring than are needed to maintain their populations, and most of the youngsters fail to survive. In addition, the youngsters always differ slightly from their parents, and some of them will clearly be more 'efficient' than others. This increased efficiency may show itself in greater size or strength, in greater resistance to disease, in better camouflage, or in some other way. Whatever form it takes, it gives the organisms an advantage in the struggle to find food and to escape from enemies. The more efficient organisms thus have a greater chance of surviving and rearing a new generation. The advantageous features are handed on to this new generation, but there is still variation and the whole process operates again. Predators and other environmental factors continue to weed out the less fit or less efficient individuals and so, generation by generation, the plants and animals improve. Because less efficient individuals are quickly removed, evolution is progressive rather than retrogressive.

EVOLUTION IN ACTION

The European peppered moth has given biologists a chance to watch evolution in action over the last 100 years. It exists in two main forms – speckled and black. Only the speckled form was known until about 1850, when a black one was found. Normally, the black one would have been caught by a bird because it would have been too obvious when sitting on a tree trunk, but the Industrial Revolution was then under way and in industrial areas trees were being blackened with soot. The black peppered moths had an advantage over the lighter ones, and began to multiply. They are now much more common than the lighter forms, although, now that smoke-abatement laws are enforced, the pale forms are beginning to gain ground.

The Birth of a Species

New species can arise in several ways, but they require some form of isolation during their development. Suppose a population of animals becomes separated into two parts by a mountain range or some other barrier which prevents the two sections from mixing and inter-breeding. Variations arising in one section might not be the same as those arising in the other, and the conditions in the two areas might also differ. Natural selection would get to work, favouring the most efficient variations in each region, and the two sections of the original population would gradually develop different characteristics. Eventually, after many generations, the two groups would be so different that they could not inter-breed even if they were brought together again. The two groups would then have evolved into two separate species.

The driving force which controls the direction of evolution may be identified solely with the environment and the other animals in it. This is why Darwin called the process *natural* selection. But Darwin must not be given all the credit for inventing the theory of natural selection. Alfred Russel Wallace, working quite independently, came to the same conclusions as Darwin at about the same time.

P. floribunda

P. verticillata

P. kewensis

The origin of a new plant species. Primula kewensis came into existence when the chromosome number doubled in a sterile hybrid between the other two species.

Genetics and Evolution

Although Darwin realized that variations were produced among plants and animals, and that at least some of them could be inherited, he did not know what caused the variations. We now know what causes them, and how they are inherited, and our newer discoveries fully support Darwin's ideas on natural selection and evolution.

Each normal cell in the body of a plant or animal contains two sets of minute thread-like structures called *chromosomes*. One set is derived from the male parent and the other set from the female parent. The number of chromosomes in each set varies from species to species. Each chromosome carries numerous *genes*, strung out along it like beads on a string. The genes are composed of the complex material called DNA and they control the manufacture of protein in the cell. In this way they control the development of the cell and of the whole body. They ensure, for example, that a robin's egg grows into a robin and not into a wren.

New combinations of genes are always being produced in the course of reproduction, because an organism receives genes from both parents. Different offspring may inherit different combinations of genes from their parents, and so they vary a good deal. Natural selection can then get to work on this variation and select the most useful combinations. There is, however, another source of variation. The genes and chromosomes normally make exact copies of themselves for each new cell formed in the body, but things occasionally go wrong when the reproductive cells are being made. The chemical make-up of a gene may alter, or a chromosome may break and join up again in the wrong order. Such changes are called *mutations* and they are passed on to the next generation, where they cause variations from the normal state. Most of these variations are harmful and the individuals possessing them fail to survive. Some of the variations are useful, however, and the individuals possessing them survive and breed. The variations then pass to the next generation, and gradually spread through the whole population. New structures can arise as a result of mutations and completely new groups of organisms can thus come into existence.

New types of animal can also arise through a process called *paedogenesis*. This occurs when, for some reason, the conditions of life of the adult have become more hazardous than those of the larvae. It is then disadvantageous to change into the adult form. Natural selection therefore favours individuals which become sexually mature, while still retaining the appearance and mode of life of the larva. It is possible that the vertebrates evolved in this way from the tadpole-like larvae of the sea squirts (see page 70). It is also possible that the insects arose like this from a young millipede stage with only three pairs of legs.

Most forms of evolution occur very slowly, but new species *can* appear quite suddenly. This is especially true among plants, where the number of chromosomes in the cell sometimes doubles. With a different number of chromosomes, the plants are clearly different species. They are often much larger than the original ones. Many of our cultivated plants have originated in this way.

The various evolutionary processes have occurred time and time again during the earth's history and have resulted in the many forms of life which are described in the following pages. It might seem impossible that the simplest forms of life could change and improve to such an extent that they eventually gave rise to dinosaurs and even to men, but natural selection can explain it all if we remember the vast stretch of time involved – more than 3000 million years. The minutest change in each generation can give rise to immense alterations over such a period.

Left: Well over 100 bird and mammal species have become extinct in the last 100 years, as a result of man hunting them or destroying their habitats. The California condor is now facing extinction. In 1964 only 40 birds were reported alive.

EVOLUTION AND EXTINCTION

The evolutionary history of an animal or plant group may be divided into four distinct stages. The first stage witnesses the *divergence* of the animal from its ancestral stock through the appearance of new features. Having diverged, the new group establishes itself and undergoes *improvement* as natural selection works upon the new features. The group is then ready for the third stage, which is *adaptive radiation* – the process of spreading out into all the available habitats. This is very well illustrated by the reptiles. Having diverged from the amphibians, they spent some time improving themselves for life on land, and then underwent a massive adaptive radiation. Without changing the basic reptilian organization, they became adapted for life in the sea and in the air as well as on the land, and acquired an immense range of forms and habits. The reptiles then declined and most reptile groups entered the final phase in the story – *extinction*.

Organisms become extinct for a variety of reasons, but a change in the environment is usually involved. Many species become *over-specialized* – so dependent on certain foods and conditions that they cannot survive any change. They die out without leaving any descendants. Less specialized organisms can evolve in the face of change and gradually turn into new species. The original species becomes extinct, but not without leaving descendants. Competition with more efficient forms of life may also lead to the extinction of less efficient forms. Many of the strange South American mammals became extinct in this way when the more advanced North American mammals swept southwards. Extinction probably faces all groups in the end, but some species have managed to delay it for millions of years and have become 'living fossils'. They usually live in places where the *selection pressure* is low. In other words, they have had few enemies and few environmental changes to hasten their evolution.

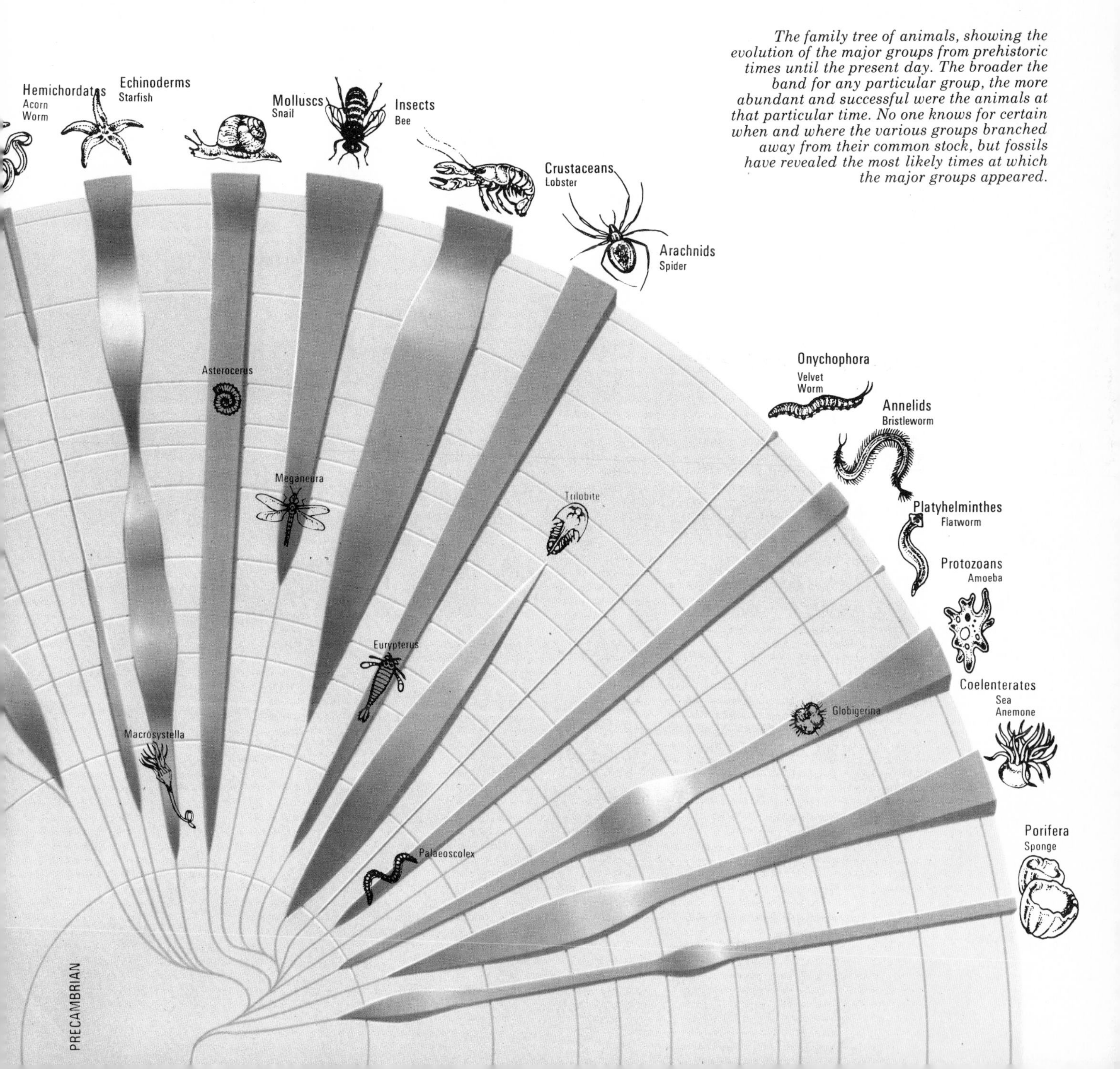

The family tree of animals, showing the evolution of the major groups from prehistoric times until the present day. The broader the band for any particular group, the more abundant and successful were the animals at that particular time. No one knows for certain when and where the various groups branched away from their common stock, but fossils have revealed the most likely times at which the major groups appeared.

The Ages of the Earth

Ever since the earth was formed, natural forces at work upon its surface have worn away the rocks, and rivers have carried the debris down to the sea. There, deep beneath the sea floor and its ever-increasing bed of sediments, pressure and heat have created new layers of rock.

This was first understood by a Scottish geologist, James Hutton (1726–97). He saw soil being stripped from the land by streams, and rivers charged with rock debris emptying into the sea. He watched cliffs crumbling as storm waves pounded away at their base; he noticed the sand of the beach being carried out to sea by undercurrents. And he concluded that the sediment being deposited on the floor of the sea was rock in the making. He realized, too, that this was part of a continuous cycle in which he could see "no vestige of a beginning, no prospect of an end". Land was uplifted from the sea and worn down by erosion, new rocks being created from the debris of the old.

The Geological Column

Millions of years of the earth's history are locked in the layers of sedimentary rock. To the geologist they are like the pages of a history book, though far more difficult to read since many are torn, turned upside down and scattered over a wide area. Unravelling the story of the rocks entails building up a complete record of the layers in their correct sequence – the geological column.

Since sedimentary rocks are formed by the deposition of rock debris, each layer is younger than the layer which was originally immediately beneath it. Even where earth movements have overturned the rock layers it is sometimes possible to tell the correct sequence from the rocks themselves. For example, if one layer of rock contains very distinctive minerals, and pebbles containing the same minerals are found in an adjoining layer, then the latter must have been created in part from the eroded debris of the former and must therefore be the younger.

The Fingerprint of Life

A greater problem in building up the geological column is that no single region contains a complete record of the past. The difficulty lies in recognizing the missing layers where they *do* occur. This is where fossils can be valuable. Because life forms have evolved continuously throughout the earth's long history, each rock formation contains a unique collection of fossils. This 'fingerprint of life' can be used to identify rocks of a similar age wherever they occur. And since the pattern of evolution is clearly recorded in rock layers which have not been greatly disturbed, it is possible to recognize rocks in one region as filling in a time gap in another region. An unnatural sequence of fossils will also show layers which have been overturned.

Two hundred million years of the earth's history are represented in the rock layers exposed in the mile-deep Grand Canyon.

Above: 'Tuning fork' graptolites (Didymograptus) are found only in certain Lower Ordovician rocks (Llanvirn series). Because graptolites were found all over the world and one distinctively-shaped species followed rapidly on another, they are invaluable in identifying strata and placing them in order.

Dichograptus

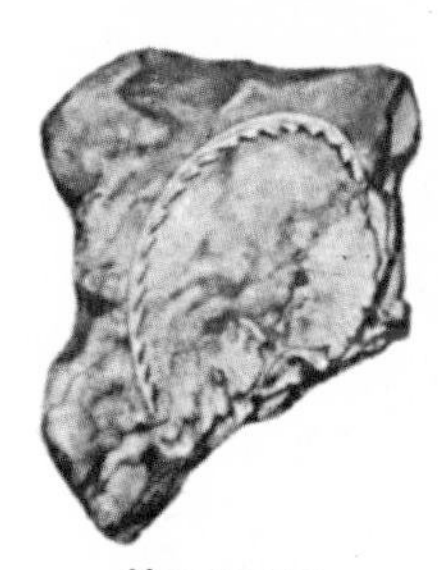

Monograptus

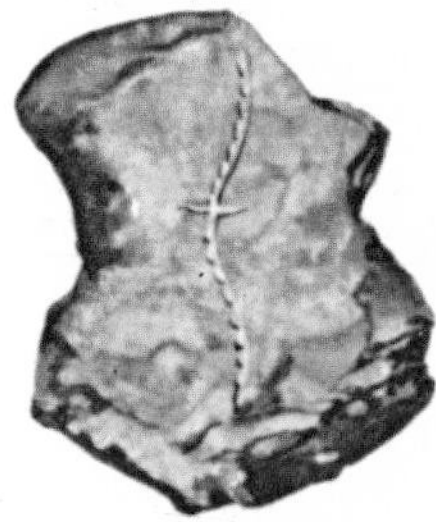

Leptograptus

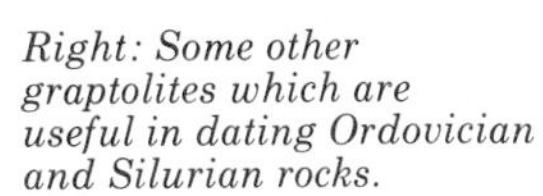

Right: Some other graptolites which are useful in dating Ordovician and Silurian rocks.

Below: Some fossils are widely distributed, while others are found only in certain places. Some are found in rocks of all ages, and some have a very restricted time range. Fossil A is widely distributed, and no use for dating rocks because it is found in rocks of all ages. Fossil B is of no use either because it is found in only one area and in many rock systems. Fossil C is very useful, however. It occurs everywhere, and a geologist finding it in a rock will know the age of the rock because the fossil was confined to a certain period of time.

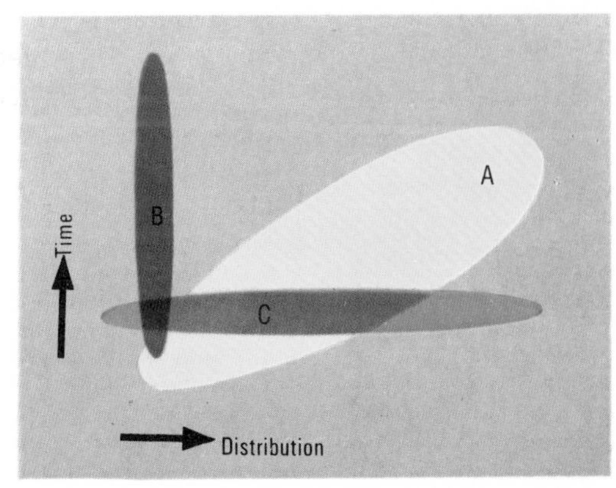

Above: Beds of Triassic rock lying on top of tilted and eroded beds of Old Red Sandstone (Devonian) in Somerset. An abrupt junction of this kind is called an unconformity and it indicates that there is a break in the geological column. The rocks missing here are those of the Carboniferous and Permian systems. During these two periods, the Devonian rocks were obviously lifted up and eroded. The Triassic rocks were later laid down on the eroded surface. Not all unconformities are as obvious as this one, but the fossils will always show if any major rock layer is missing.

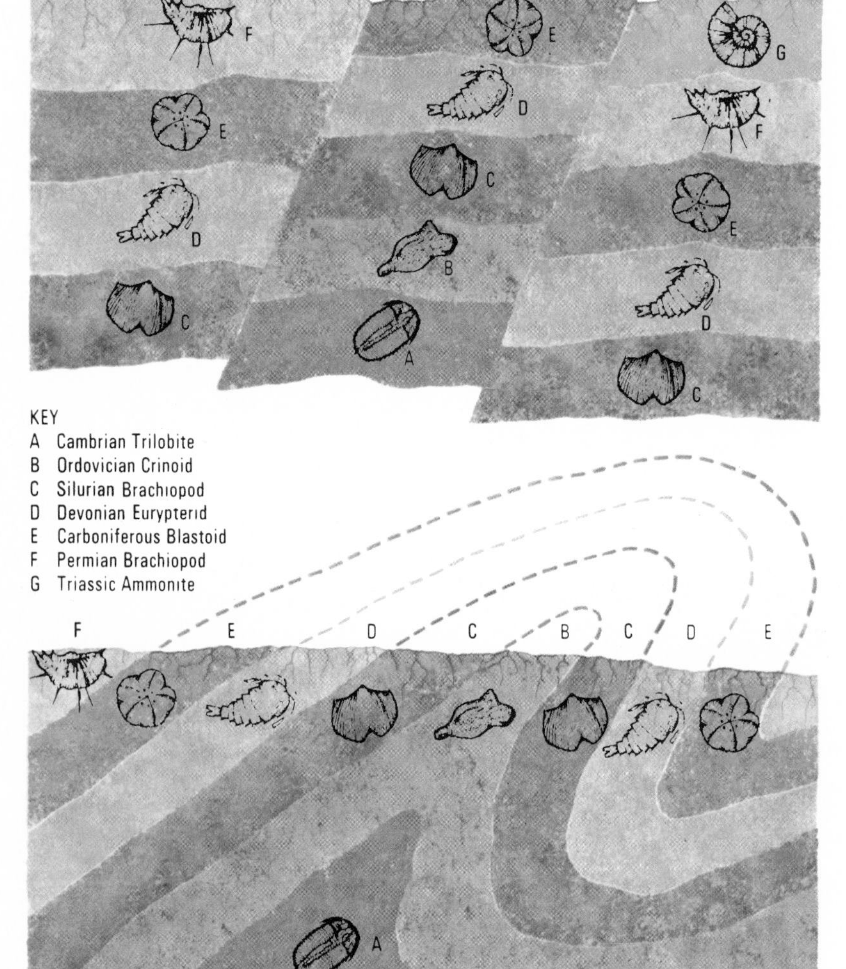

Left: Faults, where blocks of rock have slipped against each other, lead to abrupt changes in fossils at the surface. The block on the right of the diagram has remained stationary, while the other two blocks have moved up along the fault lines. The beds have thus been displaced.

Left below: An unnatural succession of fossils at the surface often indicates that the rocks have been folded and possibly turned upside down. The sequence of fossils shown in this diagram could occur only if the rocks had been lifted and folded as shown by the dotted lines and then worn down by erosion.

The geological column and the time scale it represents give the *relative* ages of geological events. It shows that the swampy forests of the Carboniferous Period flourished after the Old Red Sandstone of the Devonian Period had been laid down. But it does not give the *absolute* age of either.

The Story Unfolds

Since the geological history of the earth covers such a great span of time it is convenient to divide it into smaller units. One dividing line can quickly be drawn at the point where fossils start to appear in considerable numbers in rocks. From this time onwards, they reveal the changing pattern of life up to the present day. The two sections resulting from this division have been named the Cryptozoic Eon (Greek *kryptos* = 'hidden' and *zoe* = 'life') and the Phanerozoic Eon (Green *phaneros* = 'evident' and *zoe* = 'life'). The Phanerozoic Eon is in turn divided into three *eras* (divisions based upon life forms), namely the Palaeozoic Era (Greek *palaios* = 'ancient'), the Mesozoic Era (Greek *mesos* = 'medium'), and the Cenozoic Era (Greek *kainos* = 'recent'). The Cryptozoic Eon embraces over 80% of geological time, but any attempt to divide it into smaller units can only be carried out on a local basis, for some rocks are conspicuously poor in fossils and offer only scanty information. Eras of time correspond to *groups* of rocks, but the rocks of the Cryptozoic Eon are collectively termed *Precambrian.*

The eras of the Phanerozoic Eon are further subdivided into *periods* of time, each of which corresponds to a *system* of rocks. Both the period and the system bear the same name, which usually refers to the region where the system was first defined. Thus the Cambrian Period is named after Cambria, the Roman name for Wales, where this system of rocks was first recognized. On the other hand, the Cretaceous Period was so named because of the prominence of chalk among the rocks of this system (Latin *creta* = 'chalk').

Periods of time can be further divided into *epochs*, corresponding to *series* of rocks. In general the divisions of a period have a local rather than a world-wide importance. For instance, in Europe the Ordovician system is divided into four (or five) series. But the same system in North America is divided into just three series. Series of rock can themselves be divided into *formations.* A coal seam, for example, is a formation.

JAMES HUTTON

James Hutton (1726–1797) was a Scottish geologist who attempted to trace the origin of the various rocks in order to arrive at an understanding of the earth's history. At a time when many eminent geologists believed that all rocks had been precipitated from a primaeval universal ocean, with granite and similar rocks first and sedimentary rocks last, Hutton discovered granite veins penetrating sedimentary rocks. He concluded that the granite must have been not only molten at the time but also *younger* than the sedimentary rock.

Of equal importance was Hutton's discovery of the process leading to the formation of sedimentary rocks. He realized that the sediment deposited on the floor of the sea by streams and rivers was rock in the making. He realized, too, that this was part of an unending cycle – land is uplifted from the sea and worn down by the tools of erosion, new rocks being formed from the debris of the old. Hutton's revolutionary ideas appeared in his book *The Theory of the Earth*, which was published two years before his death.

NEPTUNE OR VULCAN?

Late in the 18th century, a fierce dispute arose between two groups of geologists, the Neptunists and the Vulcanists, over the origin and nature of rocks. The Neptunists, taking their name from the Roman god of the sea, held that the earliest rocks were precipitated out of an ancient ocean which covered the entire world. They believed that granite and basalt were the first rocks to be deposited and sedimentary rocks the last, with 'transitional' rocks, such as shales and slates, in between. But the Neptunists could neither explain how the ancient ocean had shrunk to its present size, nor how volcanoes came into being. These were believed to be burning mountains, and Neptunists even went so far as to look for signs of coal to explain their heat and fire.

The Vulcanists, taking their name from the Roman god of fire, believed that the earliest rocks were disgorged from volcanoes. Support for their position came from French geologists who were already linking basalt formations with extinct volcanoes. James Hutton's work also showed that granite intrusions into other rock layers could have occurred only if the granite was molten and had flowed into fissures before cooling.

A diagram showing how veins of granite, pushed up by immense forces within the earth, can invade sedimentary rocks. The veins can be dated by means of their radioactivity, and so geologists can get some idea of the ages of the sedimentary rocks. The latter are obviously older than the granites which have invaded them.

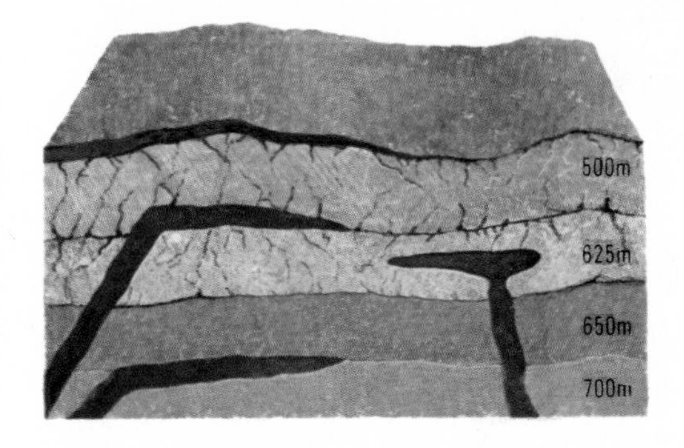

GEOLOGICAL TIME SCALE

ERA	PERIOD/SYSTEM*	EPOCH/SERIES†	YEARS AGO (MILLIONS)	CLIMATE	MAJOR GEOLOGICAL EVENTS AND ROCKS
CENOZOIC	QUATERNARY	Recent (Holocene)	0.01	Rather cool with long glacial periods. Large areas of northern hemisphere covered with ice	Sea level fell because so much water was locked up in glaciers. The land rose in many places, causing rivers to cut down again. Grand Canyon formed. Vast sheets of boulder clay deposited by glaciers. Ice finally retreated about 11,000 years ago
		Pleistocene	1.8		
	TERTIARY	Pliocene	5	Cooling down all over the world: tropical forests converted to grassland in many places	Uplift of the land continued, raising mountains even higher and causing rivers to cut more deeply into their beds. Raised beaches formed around coasts where they were lifted clear of the sea
		Miocene	25	Continuing warm	Volcanic activity and uplift increased, ending in the Alpine Revolution at the end of the epoch. The Alps and Himalayas were formed, and the Rockies and Appalachians were lifted up yet again
		Oligocene	40	Continuing warm	The land continued to rise, and the sea withdrew from large areas. Extensive lake and estuary deposits formed. More lava flows
		Eocene	55	Rather warm in Europe and North America	Uplift of the land continued with more volcanic activity. Final break between Greenland and Europe, and between Australia and Antarctica. India collides with Asia. Disappearance of Turgai Sea east of Ural Mountains
		Paleocene	65	Becoming cooler, especially near the poles. Temperate in Europe	Slow-down of continental drift and withdrawal of seas from much of the continents
MESOZOIC	CRETACEOUS		135	Warm, with rainfall similar to that of today	Continued continental drift activity and higher sea levels. Beginning of North and South Atlantic. Isolation of Africa, Australia and South America. The Laramide Revolution pushed up the Rockies and Andes at the end of the period
	JURASSIC		200	Warm, sometimes as far north as Alaska. Becoming increasingly humid as seas spread over land	Renewed continental drift activity caused sea to spread over land and deposit thick clays and limestones. Formation of Central Atlantic. Some volcanic activity and mountain-building in western North America
	TRIASSIC		225	Mainly warm and dry. Extensive deserts	Extensive salt and gypsum deposits formed in drying seas. Red sandstones deposited on land from eroded mountains. Much volcanic activity in North America
PALAEOZOIC	PERMIAN		280	Mainly warm and dry, but Ice Age continued in southern hemisphere	Meeting of Euramerica with Asia and Gondwanaland to form single world supercontinent. This caused extensive mountain-building – the Appalachians were further uplifted, together with the Urals and Austrian Alps
	CARBONIFEROUS	Pennsylvanian	300	Warm and very humid in Europe and North America, with tropical swamps. Ice Age in southern hemisphere	Thick limestones deposited in Mississippian times. Coal seams in Pennsylvanian in northern hemisphere formed as a result of continual marine flooding of the forests
		Mississippian	345		
	DEVONIAN		395	Warm and often very dry	Continued uplift of land in N.W. Europe produced the Old Red Sandstone Continent, in which great thicknesses of red sandstone were deposited. Marine deposits in North America
	SILURIAN		440	Becoming warmer: desert conditions in some places	Meeting of Europe and North America at the end of the Silurian caused the Caledonian Mountains to rise in Newfoundland and N.W. Europe. Extensive salt deposits in inland seas of North America
	ORDOVICIAN		500	Temperatures about normal, becoming warmer	Seas still widespread. More sediment filled troughs. Much volcanic activity produced beds of ash and lava. Appalachians rose up
	CAMBRIAN		600	Cold at first, with arctic conditions in North America, but gradually becoming normal	Widespread shallow seas. Sea bed sank in many places and troughs filled with sediment. Troughs started to buckle up later
PRECAMBRIAN TIME Time extending back to the formation of the earth			4600+	Atmosphere with little oxygen, especially in early stages. Often cold, with glacial periods. Some warmer periods	Great thicknesses of sedimentary rocks accumulated, but most of those surviving have been greatly altered by folding and pressure. Much volcanic activity

**Period refers to time and System to rocks.*

†Epoch refers to time, and Series to rocks

MOUNTAIN BUILDING	MAJOR FEATURES OF PLANT LIFE	MAJOR FEATURES OF ANIMAL LIFE	PERIOD/SYSTEM
	Development of arctic floras able to withstand harsh climates. These floras later left as relict floras on tops of mountains when glaciers retreated	Java and Peking Man evolved from Australopithecus type and eventually gave rise to modern man. Woolly rhinoceroses and mammoths coped with arctic conditions	QUATERNARY
	Development of modern temperate floras as climate cooled down. Coniferous forests in north. Further spread of grasslands around tropics	Ape-man (Australopithecus) in Africa towards end of epoch. Numerous kinds of elephants and other large mammals, but most of the larger ones died out as the climate cooled	TERTIARY
	Grasslands began to appear, and rapidly increased	Numerous apes in Africa. The ancestors of modern man diverged from the apes. Herds of grazing animals appearing on the grasslands	
	Continued increase in flowering plants	Early apes appeared. Many other modern mammals beginning to evolve	
	Plants of very modern appearance were spreading all over the world. Flowering trees dominated large areas	Many strange herbivorous mammals appeared. Early horses and elephants	
	Rapid increase in flowering plants. Cycads declined	Mammals evolved rapidly and filled niches left by reptiles	
	First appearance of flowering plants. Conifers, ferns and cycads still common	Dinosaurs and pterosaurs continued to dominate the land at first, but became extinct later. Many birds and small mammals. Extinction of ammonites and many other marine creatures	CRETACEOUS
	Cycads abundant. Maidenhair trees (Ginkgo) common. Widespread thickets of large ferns	Ammonites abundant in sea, together with new kinds of coral and sea urchins. Dinosaurs abundant on land. Flying reptiles (pterosaurs) and the first birds. Small mammals	JURASSIC
	First appearance of cycads and Bennettitaleans. Conifers continued to dominate the land	Continued evolution of reptiles produced the first dinosaurs and the large marine reptiles. End of the mammal-like reptiles; evolution of the first mammals. Ammonites flourished in sea	TRIASSIC
	Giant clubmosses and horsetails became extinct in the face of drier conditions. Smaller seed ferns replaced large types. Great increase in conifers.	Rapid increase and spread of reptiles. Amphibians less important and mostly in the water. Widespread extinctions in the sea: trilobites became extinct, along with many fish and corals	PERMIAN
	Giant clubmosses, ferns and horsetails in the coal swamps of northern hemisphere. Seed ferns and first conifers. Glossopteris flora in southern hemisphere	Amphibians increased and spread; soon gave rise to reptiles. Insects appeared and became very common. Corals, brachiopods and fishes very common in the seas	CARBONIFEROUS
	Land plants became common and widespread. Several new types appeared, including ferns, clubmosses, and the earliest seed plants	The age of fishes – many kinds in fresh and salt water. Graptolites died out. Goniatites – ancestors of ammonites – appeared. Amphibians evolved from fishes and moved on to land	DEVONIAN
	Algae continued to thrive in the water. The first known land plants appeared – leafless and rootless plants known as Cooksonia	Graptolites decreased. Jawless, armoured fishes abundant. First jawed fishes appeared. Large sea scorpions (eurypterids). Coral reefs and abundant brachiopods	SILURIAN
	Lime-secreting algae abundant, often forming small reefs. No known land plants, although simple mosses and lichens may have lived on land	Graptolites and trilobites abundant. Corals and brachiopods spread rapidly. Earliest vertebrates – armoured fishes	ORDOVICIAN
	Lime-secreting algae contributed to rock formation in Upper Cambrian	Fossils abundant in rocks. All major groups of invertebrates represented. Graptolites, primitive shellfish, corals, echinoderms, crustaceans and other arthropods. Trilobites especially common	CAMBRIAN
	Algae, fungi and bacteria. Blue-green algae well preserved in ancient cherts	Rare traces of animal life in later Precambrian rocks – less than 1000 million years old. Forerunners of trilobites, worms and jellyfishes.	PRECAMBRIAN

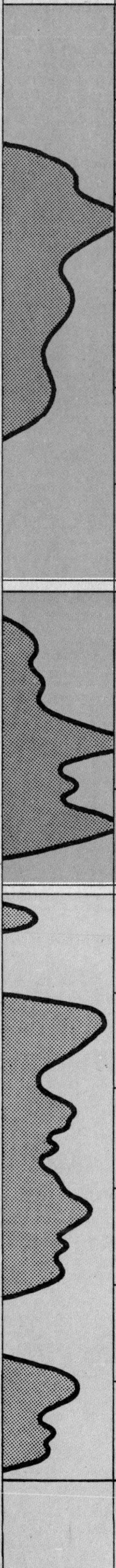

Sabre-toothed tiger from the Pleistocene.

An ammonite from the Jurassic seas.

Annularia, a plant fossil from the Carboniferous.

A trilobite from the early Palaeozoic Era.

The Birth of Life

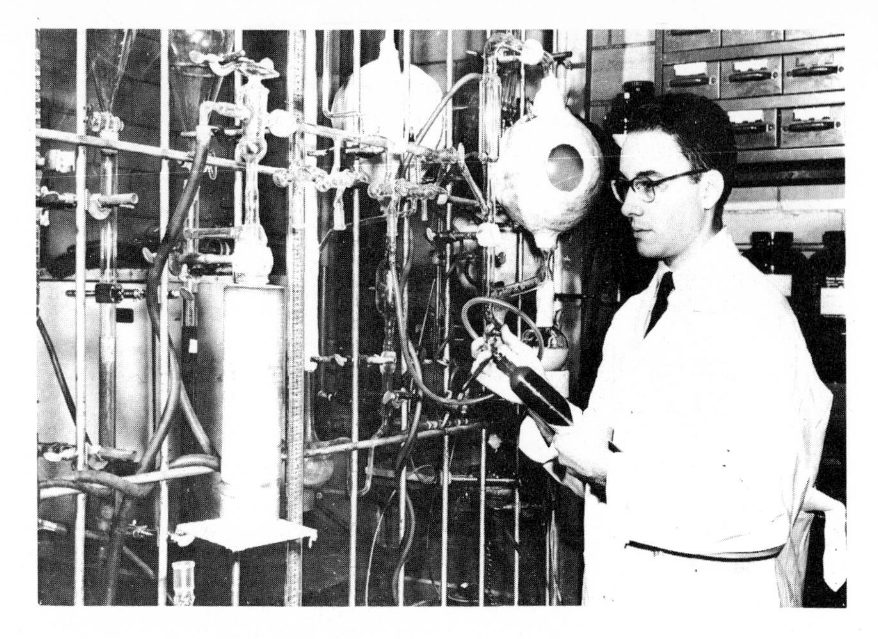

For much of the early history of the earth, both the land and the sea were lifeless. The atmosphere contained a smothering mixture of gases such as ammonia, methane, hydrogen, carbon dioxide and water vapour. Radioactivity, lightning, volcanic upheavals and the rays of the sun provided a continuous source of energy to fuel an unending chemical activity. Atoms and molecules were shuffled about; new chemicals were formed. Carbon, which has the ability to make long, complex molecules, was worked into countless different forms. These carbon, or organic, compounds, became the building blocks of life.

Above: Dr Stanley Miller, an American scientist, carried out experiments which support the idea that the first amino-acids could have been formed from the ammonia, methane and hydrogen in the earth's primitive atmosphere. Miller sent sparks (lightning) through a mixture of these gases in steam and he found organic compounds in the water afterwards.

THE NEW ATMOSPHERE

The beginning of life profoundly transformed the composition of the atmosphere. Like microscopic factories, the photosynthesizing plants steadily consumed the carbon dioxide in the air while pumping back oxygen as a waste product. High up in the atmosphere, some of the oxygen was transformed into a protective ozone shield which the grilling solar radiation could not penetrate. With the air rich in oxygen, the scene was set for plants and animals to begin their conquest of the earth.

Nature's Test Tube

Carbon compounds were washed into the seas from the atmosphere, where they accumulated along with other minerals to form a rich chemical 'soup'. This was to be nature's test tube, where experiments took place in building the giant molecules that could create and maintain life. Here, in particular, a variety of amino-acids accumulated, and from these the first proteins were built.

The formation of proteins is an essential step in creating life. At first it occurred in a completely random way. But nature had endless reserves of time – the most important ingredient of all. Hundreds of millions of years went by, until eventually a cycle evolved in which nucleic acids were able not only to string amino-acids together to form proteins but also to make more nucleic acids.

The first protein cycles were probably very fragile, and readily broken down. But eventually a great advance took place: some protein cycles developed a protective wrapping, or 'skin', of fatty material. This would have freed them from unwanted chemicals which might have poisoned the protein cycle, and would thus have made them more stable; it would also have held together the materials inside. Such skins might have developed from the oily film on the surface of the sea, when it had been broken up by the waves into individual droplets.

Below: A simplified diagram of Dr Miller's spark discharge apparatus.

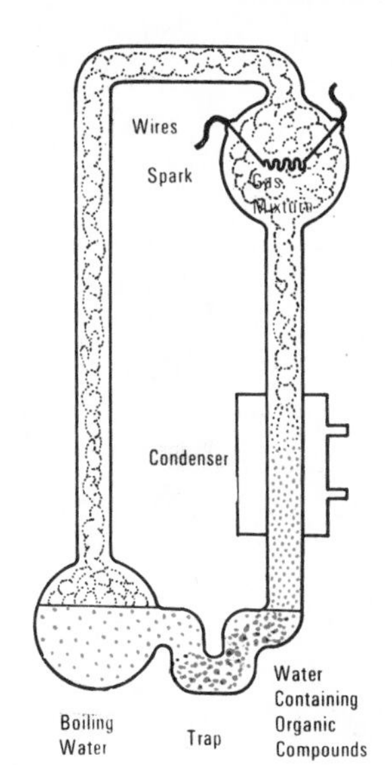

In this way, the first primitive cells were born. Cells which contained the molecule DNA were able to produce copies of themselves. Those which were most like the parent cell continued to survive, while freak cells rapidly died out and disappeared.

Even over millions of years of leisurely experimentation, life would not have been created by this haphazard method alone. The laws of natural selection (see page 27) were already at work. It was always the most stable and hardy nucleic acids and proteins which tended to predominate in the chemical soup.

Harnessing the Sun

The first living microscopic organisms were bacteria which fed on organic chemicals. These they extracted from the soup and fermented to provide themselves with the energy to live and grow. As the chemical soup was limited in quantity, these bacteria seemed destined to eat themselves out of existence. But some cells were able to break out of the trap by learning how to make use of sunlight. Sunlight was a direct source of energy which they could harness, with the help of a complex molecule called chlorophyll, to produce their own food. This process, called photosynthesis, was the important breakthrough which permitted the development of all life as we know it. The new cells were the first primitive plants. They were able to grow rapidly and efficiently, for they were no longer dependent on chemicals formed at random to supply all their needs. Other microscopic cells also evolved which could use the oxygen that was being produced as a by-product of photosynthesis. Some of these cells developed a whip shape. They were able to move around with a thrashing movement in their search for food particles. These simple, sea-dwelling microbes were the only forms of life for more than half the earth's history.

Precambrian Life

The Precambrian rocks were once thought to be devoid of fossils apart from the curious cabbage-like masses called stromatolites, and the nature of these was in doubt for a long time. But during the 1950s other kinds of fossils came to light which showed that living organisms existed on the earth at least 3100 million years ago. These ancient fossils are mainly of bacteria and, considering how small and fragile these early organisms were, it is amazing that they were preserved at all. It is certainly not surprising that they lay undetected for so long.

All living organisms, apart from the viruses, belong to one of two major divisions. The simpler of the two groups contains the *prokaryotes* – organisms whose cells contain no nuclei and only a single strand of genetic material. This

SPONTANEOUS CREATION

For many years, people believed that life came into existence spontaneously. Worms were thought to be born in mud, maggots in rotting meat, and mice in mildewing hay. Though these examples were easily proved wrong it was only in the mid-19th century that Louis Pasteur was able to prove that thoroughly sterilized materials could not give rise to living things.

In this experiment, water dropped on to molten lava has resulted in the formation of organic molecules. This is one way in which life could have originated.

Above: An ancient stromatolite which has been sliced through to show the layered structure.

group contains only the bacteria and the blue-green algae. All other plants and all the animals belong to the second group, known as the *eukaryotes*. Their cells all have nuclei and their genetic material is all properly organized into chromosomes (see page 29). Not surprisingly, perhaps, the oldest known organisms are prokaryotes: no eukaryotes are known from rocks more than about 1000 million years old.

The Oldest Organisms

The oldest undoubted fossils, about 3100 million years old, belong to the Fig-Tree Cherts of Rhodesia, one of the most famous of all geological strata. Odd structures are found in some even older Rhodesian rocks, but it is doubtful if these are fossils. The Fig-Tree Cherts are black, glassy rocks composed almost entirely of silica – the same material that makes flint and sand. They apparently accumulated on the floor of a lake or a shallow sea where, by processes not yet understood, the silica formed a jelly-like deposit. As this material hardened to form the cherts, it trapped the organisms which had been living on the bottom. These died, but they left natural moulds in the silica and some left their cell walls as well. The fossils are very tiny, but they are perfectly preserved and, when the rocks are cut and polished and examined with an electron microscope, they show

Below: Blue-green algae from the Fig-Tree Cherts, magnified more than 1000 times.

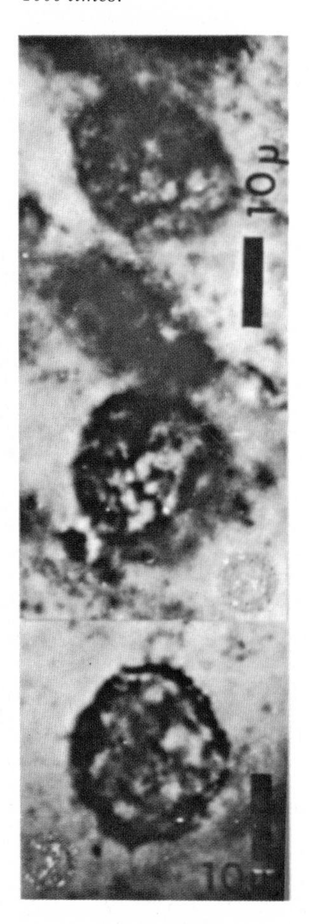

up as little rods, spheres and filaments. Micro-organisms of similar appearance are abundant today. And because of their resemblances, the small rods are interpreted as bacteria and the spheres as blue-green algae.

Minute traces of organic chemicals extracted from the Fig-Tree rocks include carbon compounds called pristanes and phytanes, which nowadays form during the decomposition of chlorophyll. This is a strong indication that chlorophyll was present at this early date. It is quite likely, therefore, that photosynthesis was already taking place, releasing free oxygen into the primitive atmosphere and gradually bringing about the changes that led to the present-day composition of the air. The bacteria may have been photosynthetic as well as the algae: alternatively, they may have obtained their energy by processes other than photosynthesis, as do most of today's bacteria.

The next oldest microfossils are only about 2000 million years old. These come from the Gunflint Cherts on the northern shore of Lake Superior in Canada, and

STROMATOLITES

The extraordinary cauliflower-like masses of the Precambrian stromatolites have now been shown to have been formed by blue-green algae. They consist of layer upon layer of lime, secreted by the algae or else trapped by them among their sticky filaments. The most ancient stromatolites are perhaps 2800 million years old, and come from Africa.

Stromatolites are common enough in ancient rocks, but it was not until living examples were found that they could be shown definitely to have resulted from the growth of blue-green algae. Such modern stromatolites grow in Shark Bay in Western Australia, where they form large flat-topped pillars.

Stromatolites must have been one of the dominant and certainly the largest of the life forms of the Precambrian: they were even capable of forming reef-like structures. They declined at the end of the Precambrian and far fewer are found in later rocks. This decline may have been due to the rise of animals such as snails, which today feed on blue-green algae wherever they can find them. The water of Shark Bay is too salty for snails, but not for the tolerant blue-green algae. Stromatolites can therefore live there without hindrance.

Opposite: A blue-green alga from the Gunflint Cherts of North America, very highly magnified.

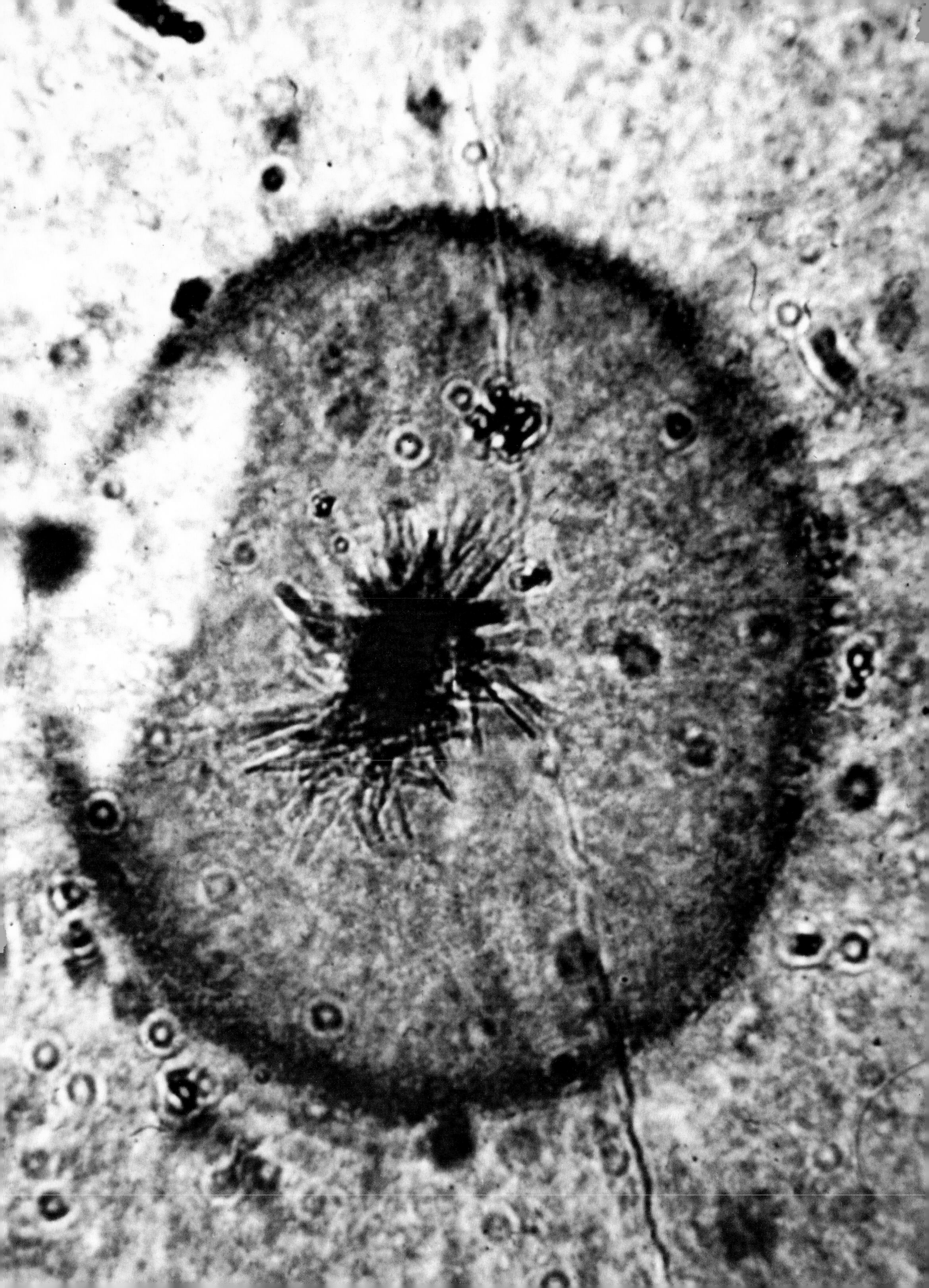

from similar beds elsewhere. The Gunflint Cherts are curiously banded rocks in which layers of chert alternate with layers of ironstone or form stromatolitic masses rich in iron. The fossils are most numerous in the chert bands, which may have been formed in the same way as the Fig-Tree Cherts, the tiny organisms being entombed in a rapidly-setting silica jelly. All of the fossilized organisms are prokaryotes, but they are much more diverse than those of the Fig-Tree Cherts. There are many different kinds of blue-green algae, mainly spheres but also including some jointed and unjointed filaments which probably formed mats on the sea floor. Some of them have quite an elaborate structure, perhaps the most peculiar being an alga called *Kakabekia*, which has a basal bulb attached by a short stalk to a ribbed canopy like an umbrella.

Many of the Gunflint fossils are similar to modern blue-green algae. Some *Kakabekia*-like algae have even turned up in the ammonia-rich sands near the mediaeval sewage outlet of Harlech Castle in North Wales. The Precambrian *Kakabekia* flourished in an environment where there was still quite a lot of ammonia in the atmosphere and dissolved in the water, and it seems that the modern form may have the same requirements. Modern blue-green algae are amazingly tolerant of harsh environments. Some can live in springs of near-boiling water; others are unaffected by strong ultra-violet radiation. These modern blue-green algae have obviously retained all the toughness and versatility which enabled them to survive in the grim conditions of the Precambrian. But they have evolved very slowly, for their great tolerance has meant that there has been less 'selection pressure' on them than on other creatures (see page 31) and there has thus been little change in them in the course of hundreds of millions of years. In any case, being prokaryotes, the blue-green algae do not have the necessary genetic equipment for rapid evolution: only cells with proper chromosomes can evolve quickly.

Because Precambrian fossils are so few and scattered in time, we cannot be sure when the first eukaryotes appeared, but eukaryotic algae definitely exist in the Bitter Springs Formation of Australia. This is a chert deposit which was formed

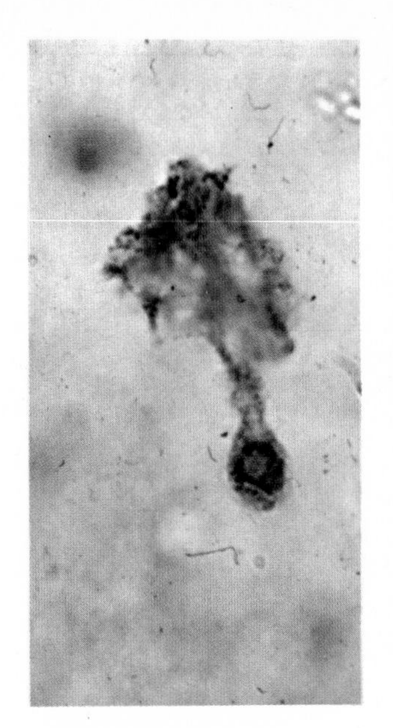

Above: Kakabekia, an alga from the Gunflint Cherts. The 'umbrella' and basal bulb can be clearly seen.

ALGAE

The algae are a very varied group of predominantly aquatic plants. They range from the familiar seaweeds to minute one-celled types such as diatoms, which float in water and are an important component of plankton. Algae are split into a number of groups according to their colour (e.g. green, blue-green, red and brown). These colours are produced by a combination of pigments. In green algae, as in land plants, the dominant pigment is chlorophyll.

The earliest blue-green algae were unicellular (that is, the whole plant comprised just one cell). But in younger rocks, these unicells are linked to form chains (filaments) and joined up into masses (colonies). The multicellular types, too, are almost identical with certain living forms.

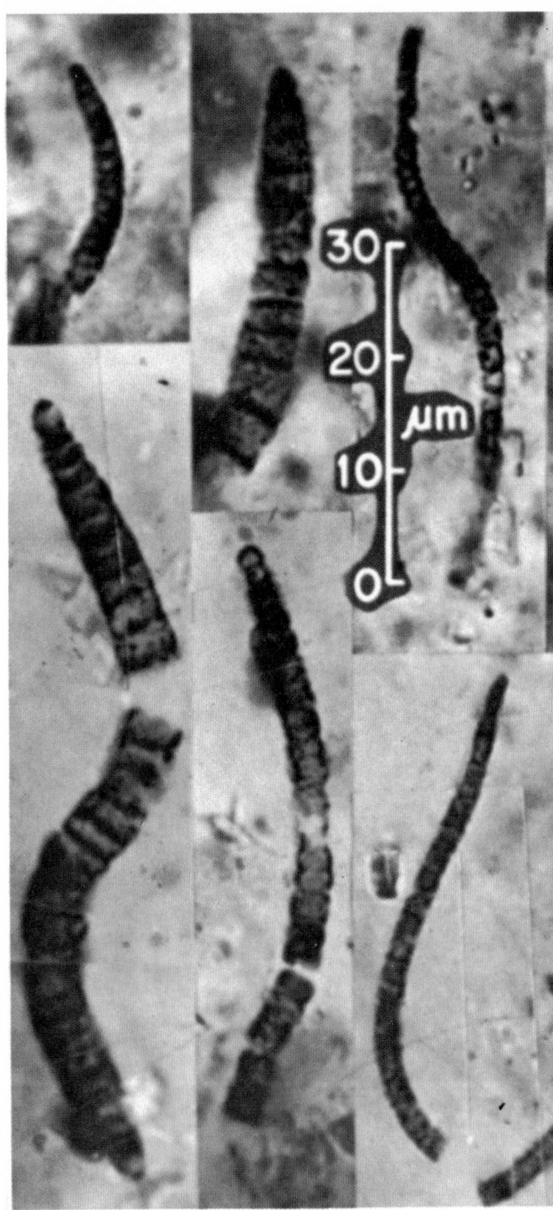

Right: Some filamentous algae from the Bitter Springs Formation in Australia, laid down about 1000 million years ago. Some of the algae are represented only by hollow moulds in the chert, but the organic material of some of them has been preserved. Unlike the Fig-Tree and Gunflint fossils, some of these algae had nuclei.

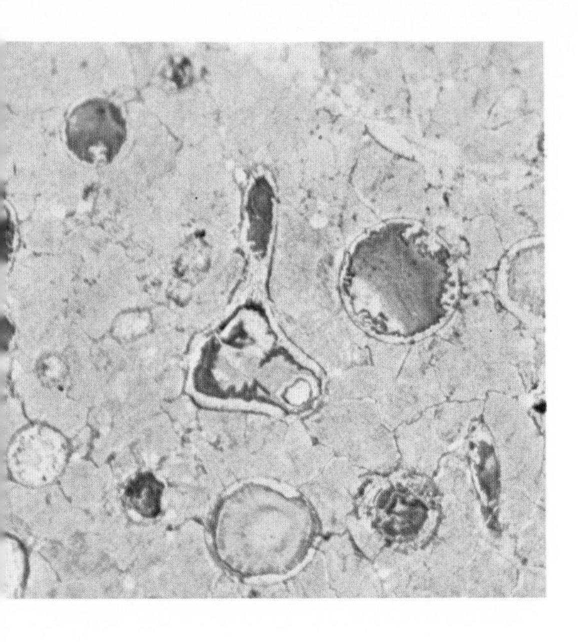

Left: About 2000 million years old, Vallenia erlingii is the best preserved of a number of organic structures known from the Ketilidian sediments of south-west Greenland.

about 1000 million years ago, and which is now almost as famous as the Fig-Tree and Gunflint Cherts. Various kinds of algae are present in the Bitter Springs rocks. Blue-green algae are abundant, but there is also a new genus of algae, known as *Glenobotrydion*, in which the nuclei are actually preserved. As it has nuclei, *Glenobotrydion* must be a eukaryote. It is very likely, however, that the origin of the eukaryotes belongs further back in time. Some micro-organisms have very recently been reported from Australian rocks about 1600 million years old and, although no nuclei have been found, other features suggest that these organisms may be eukaryotes.

The Oldest Animals

The rise of the eukaryotes was a major biological change in itself, but it also provided the impetus for much more rapid evolution. Within 300 million years from the appearance of *Glenobotrydion* we see the first real animal fossils, and these are already so well formed that the origin of the first animals must be dated decidedly earlier. The Ediacara Fauna of Central Australia, discovered in 1947, is over 700 million years old and contains a fine array of many kinds of real animals. There are creatures like modern sea pens, various sorts of jellyfish, and beautiful segmented worms, as well as certain oddities which do not seem to fit into any known category.

The Precambrian fossil record is full of gaps, but they are slowly being filled. Very recently there has come to light an assemblage of bizarre fossils from the Namib Desert of South West Africa. These are older than the Ediacara fossils and very difficult to interpret. Some branch like corals, others have a segmented filamentous structure, while others again are egg-shaped with a curious surface sculpture. It is not really known what these are. They do not closely resemble any known animals or plants and it has even been suggested that they are some kind of organism intermediate between the animal and plant kingdoms. None of them seems to have been capable of movement, however, and it was only when highly mobile animals first evolved that the real explosion of life could begin.

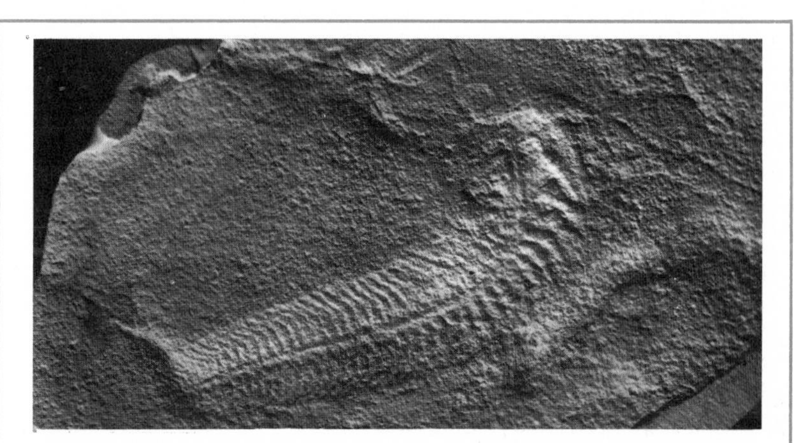

Above: Spriggina, a worm from the Ediacara Fauna. Moulds of its tough coat were left in the rocks.

Below: Some members of the Ediacara Fauna as they must have appeared in life. Jellyfishes such as Kimbrella (A) and Ediacara (B) were particularly common. Sea pens (C) sprouted from the sea bed, while Dickinsonia (D) and other worms crawled over the mud. Cyclomedusa (E) was a primitive bottom-dwelling relative of the jellyfishes, and Tribrachidium (F) was a forerunner of the starfishes and sea urchins.

The Progression of Plants

Right top: Fossilized Solenopora, one of the lime-secreting red algae which helped to form reefs in the Palaeozoic Era.

We saw in the previous chapter that the blue-green algae were among the earliest of living organisms. They were, in fact, the dominant organisms for about two-thirds of the time during which plants and animals have been on the earth. We know a lot about these early algae because some of them were perfectly preserved in rocks called cherts. But our information on the later history of the algae is largely restricted to those groups whose soft tissues were supported by skeletons of silica or lime.

Calcareous algae were important rock builders during Precambrian and Palaeozoic times. They built up rocks with their own skeletons, or else they trapped and bound limestone particles drifting in the water. These calcareous algae included representatives of the red, green and blue-green algae and were very similar to those living today. Somewhere among these early algae, probably among the green algae, were the ancestors of the land plants.

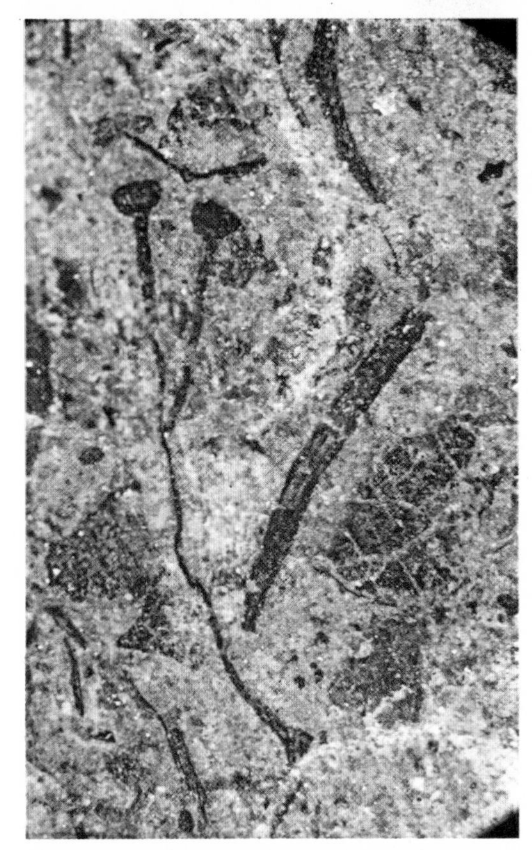

Right centre: A fossil of Cooksonia in Silurian rocks from Pembrokeshire (magnified 12 times), and a reconstruction of the plant (above) showing the forking stems and spore capsules.

The First Land Plants

One of the most striking features of the present-day landscape is the abundance of green vegetation. This is composed mainly of flowering plants, ferns and conifers. These are known as vascular plants, and all possess water-carrying tubes – an adaptation for life on land (see page 82). They first appeared late in Silurian times – just over 400 million years ago. Earlier land surfaces were either strikingly bare or perhaps clothed with a thin scum of algae, rather like the green film seen on many tree trunks today.

The earliest recorded vascular plant is called *Cooksonia*. It had no roots, leaves or flowers and was essentially a cluster

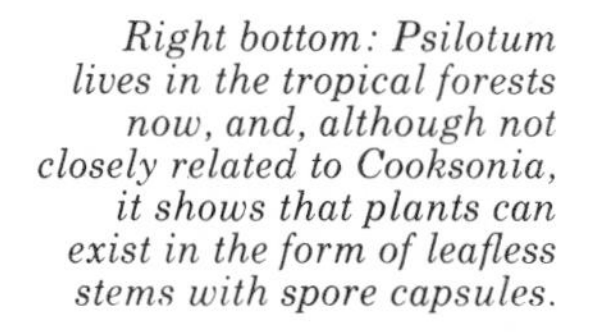

Right bottom: Psilotum lives in the tropical forests now, and, although not closely related to Cooksonia, it shows that plants can exist in the form of leafless stems with spore capsules.

of erect forking stems. Each stem ended in a spore-filled capsule or sporangium. A second type of vascular plant appeared a little later than *Cooksonia*. This was *Zosterophyllum*. Its stems were smooth and leafless like those of *Cooksonia*, but its sporangia were grouped into little spikes on the sides of the stems.

There was a very rapid increase in the number of terrestrial plant types early in the Devonian Period, and these new types colonized land areas in many parts of the world. Their fossils are found mainly in sediments which were laid down in lakes and rivers, and it seems likely that most of the plants grew near the water. *Cooksonia* and *Zosterophyllum* were still present, and many of the new plants were modifications of these two basic types. Some had prickly stems, for example, and some became so highly branched that each plant looked like a little bush. *Psilophyton* was an important plant of the latter type. None of these plants is living today, however. A present-day observer would not recognize any of the plants which lived in the Lower Devonian, except for some of the clubmosses. These were relatives of the ferns, and they hold the record for being the vascular plants with the longest history. Some of today's clubmosses look quite similar to those of Devonian times.

Above: A fossil of Zosterophyllum showing the spore capsules.

The early Devonian Period saw plants firmly established on the land. By the middle of the period the majority of the early vascular plants had become extinct, but descendants of the two original types remained in the main stream of plant evolution. Clubmosses, which are thought to have evolved from the *Zosterophyllum* group, flourished and increased in size. Plants similar to *Psilophyton* are believed to have given rise to an assortment of strange types among which can be seen the beginnings of the ferns, the horsetails, and even the seed-bearing plants. Some of the plants began to produce extra water-carrying tissues.

The tough walls of these gave the plants much greater strength. This eventually led to the evolution of the tree, a very efficient kind of plant in the competition for space.

The Evolution of the Seed

The next major event in plant evolution was the appearance of seeds. The earliest examples come from the Upper Devonian rocks of the United States and are quite unlike anything growing today. These ancient seeds were carried at the ends of stems and were protected by clusters of branching stem-like structures. Unfortunately, we know nothing about the plants which produced them.

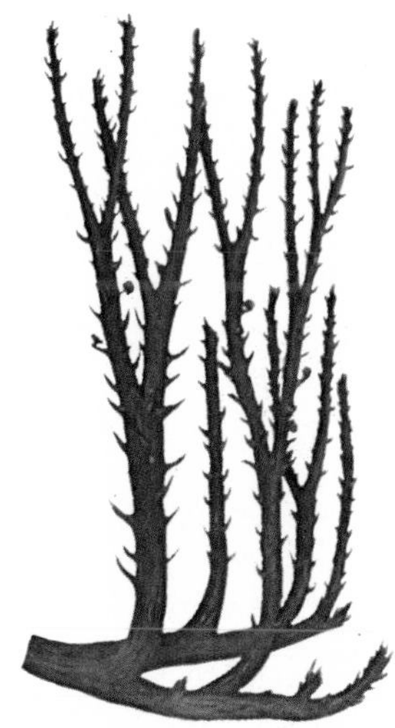

Above: Drepanophycus was a Devonian clubmoss, or lycopod, and, although much larger than today's clubmosses, its basic appearance was much the same.

Left: A modern clubmoss.

Right: Some of the 'experimental' plants of Devonian times: Psilophyton (1), Pseudosporanchus (2), Cladoxylon (3), Hyenia (4) and Aneurophyton (5). Some were related to today's ferns, while others were more closely related to the clubmosses.

Above: A 19th century illustration of fossilized tree trunks exposed in a quarry in the Swansea Valley.

Below: A diagrammatic representation of a coal forest, showing some of the major types of plants and the fossils they left behind. The dark band in the ground represents the plant debris which now forms the coal seams. The other bands are the sand and mud deposited by the flood water.

The inset fossils are as follows: Alethopteris (1), Sphenophyllum (2), Neuropteris (3), Pecopteris (4), Annularia (5) and the bark of Lepidodendron (6). (Unless they are accompanied by the remains of sporangia or seeds, it is impossible to classify fossilized fernlike leaves. They are called 'form-fossils', defined only by their shape. Alethopteris, Neuropteris and Pecopteris are form-fossils of this kind. The name Calamites really refers only to casts of the stem of this plant. The form-fossils of the leaves are called Annularia.)

We can trace the stages in the evolution of the seed by looking at some of today's plants. Ferns and clubmosses reproduce by scattering tiny spores. These spores are usually all alike, but the clubmoss *Selaginella* produces two kinds of spore, one large and one small. The prothalli (see panel) develop *inside* the spores, and those in the smaller spores then release male cells. The latter then enter the larger spores and join with the female cells to begin a new generation. *Selaginella* occasionally fails to release some of its larger spores, and the female cells inside are fertilized while still on the parent plant. Fossils show us that some of the large clubmosses and horsetails of the Carboniferous Period also retained their large spores in this way. Keeping the large spore and its prothallus attached to the parent plant was clearly useful because it provided protection. It was undoubtedly this advantage which led to the evolution of the seed. We do not usually think of the conifers and the flowering plants as producers of spores, but they do produce two kinds, just like *Selaginella*. The small ones are released as pollen grains, while the large ones are always retained on the parent plant. They are enclosed in various layers which protect and nourish them and, after fertilization by cells from the pollen grains, they become seeds. Seed-bearing plants fall into two groups – the gymnosperms, which include the conifers and their relatives, and the flowering plants, or angiosperms. Gymnosperm seeds are naked, but angiosperm seeds are always enclosed in fruits.

COMPLEX LIFE HISTORIES

The ferns have neither flowers nor seeds. They reproduce by scattering tiny spores which are produced in little capsules on the undersides of the leaves (see the photograph above). The spores are scattered by the wind, and germinate when they fall on moist ground. But they do not grow into new fern plants: each one produces a little plate-like plant called a prothallus. This is only a few millimetres across, but it is the true seat of reproduction. It produces male and female cells which pair up when they are ripe. An embryo is thus formed, and this grows into a new fern. Horsetails and some clubmosses have similar life histories, and so, presumably, did all the early vascular plants. But the prothallus is a very delicate and vulnerable part of the life cycle, and the seed plants have eliminated this free-living stage. There is still a prothallus, but it remains on the parent plant. Fertilization takes place there and the whole system – spore capsule, spore, prothallus, and embryo – makes up the seed.

Low-growing **seed ferns** overhang the water. The fronds bear tiny seeds. Behind them are the whorled stems of the horsetail ***Sphenophyllum.***

Medullosa, with its long roots reaching from the trunk.

The Carboniferous Forests

Perhaps the most familiar fossil plants are those belonging to the Carboniferous rocks, particularly the Coal Measures. About 300 million years ago much of Britain and North America was covered by freshwater swamps and lagoons. The climate was very warm, and luxuriant forests grew in the swamps. Dead and dying plants fell into the water and their remains accumulated to form thick layers of peat, like that being formed in bogs today. Changes in sea level during Carboniferous times led to flooding, which killed off the vegetation growing on the peat and deposited layers of sand and mud on top of it. The vegetation grew again when the floods eventually subsided, but then the whole process began again, to be repeated several times. The weight of the overlying mud and sand gradually converted the peat into coal. During this process of compression and compaction the structures of the individual plants were usually destroyed. Occasionally the

Lepidodendron (right), a gigantic clubmoss. At its base, the trunk is divided into four main rootstocks which then branch regularly themselves.

Calamites (left), a giant horsetail, grew from a large underground rhizome.

Psaronius a tree fern, with a layer of fibrous rootlets on the trunk.

Cordaites was a tall and much branched tree with strap-shaped leaves. Seeds hung on long stalks from loose cone-like buds. Cordaites formed large forests in Carboniferous times.

impression of a leaf can be seen in a piece of household coal, but most plant fossils are found in the shales which lie between the coal seams.

The coal swamps contained herbs, shrubs, and trees, but the trees were quite unlike those of today. They were not flowering trees or conifers, but huge clubmosses and horsetails – groups which are represented today only by small, herbaceous species. The giant clubmosses, or lycopods, were about 30 metres (100 feet) tall, with trunks up to two metres (six feet) in diameter. *Lepidodendron* is a well-known example. Although large, its trunk was much less woody than modern tree trunks and it certainly would not have been a good timber tree. The tree-like horsetails, such as *Calamites*, reached heights of about 18 metres (60 feet). Another horsetail, *Sphenophyllum*, was common in the herbaceous layer, where it scrambled among other plants.

The Carboniferous Period is often referred to as The Age of Ferns. Fern-like plants were certainly very common, but most of them were actually seed ferns – seed-bearing plants more closely related to the conifers than to true ferns. Seeds are very rarely found attached to the fossilized fronds, but the internal structure of the fronds does indicate that the plants were gymnosperms. The seed ferns were probably the dominant plants of the shrub layer.

Different kinds of vegetation existed in the drier areas further inland. Herbaceous clubmosses probably covered large areas, while gymnospermous trees grew on the hills. Most of these gymnosperms belonged to a now extinct group called the Cordaitales. Their tall, woody trunks reached heights of 30 metres (100 feet).

The last important event of the Carboniferous Period was the arrival of the conifers, whose seeds are carried on the scales of woody cones. These trees are still with us in the shape of firs, pines, spruces, and many other species. Although greatly outnumbered by the flowering trees in the world as a whole, they dominate the forests in the cooler parts of the world.

New Plants for Old

The end of the Palaeozoic Era was a time of change. The giant lycopods and horsetails, which were adapted for life in moist

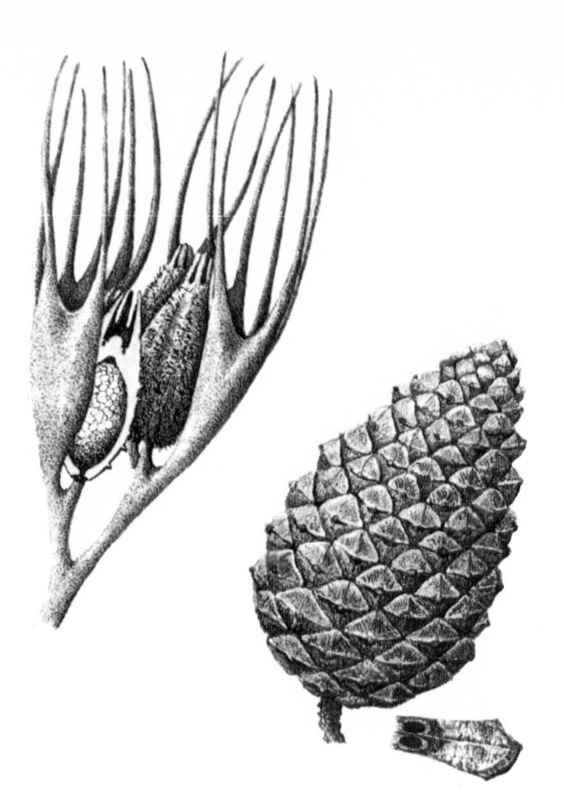

Above: Archaeosperma (left) comes from Upper Devonian rocks in North America and is the oldest known seed. It is surrounded by leafy bracts, quite unlike the woody cones (right) in which the modern conifers carry their seeds.

Below: Certain layers of the coal seams contain rounded boulders called coal balls. These consist of densely packed plant fragments impregnated with lime. Thin slices of these coal balls (below) reveal a great deal about the structure of the plants.

A Jurassic delta landscape, dominated by gymnospermous trees, such as the swamp cypress on the left and the maidenhair tree on the right. Branching cycads and bennettitaleans mingle with numerous ferns and tree ferns, while large horsetails grow in the shallow water, just as reeds and some small horsetails do today. Notice that no flowering plants have yet appeared.

conditions, became extinct in the drier climate of Permian times. *Neuropteris* and some of the other Carboniferous seed ferns lingered on into the Permian, but the age of the luxuriant, large-leaved seed ferns was ending. The newer seed ferns had much smaller and usually tougher leaves, which were better able to cope with the drier conditions. The true ferns persisted throughout this time of change and became the most important ground cover plants of the Mesozoic Era. Faced with competition from the fast-evolving conifers, the Cordaitales became extinct. Flowering plants had still not appeared, but world vegetation was otherwise taking on a distinctly modern appearance. Early Mesozoic conifers looked like modern types with cones very similar to those we see today.

The Mesozoic Vegetation

The Triassic landscapes of 220 million years ago must have been arid and bare in many places, but here and there were the beginnings of the luxuriant Jurassic vegetation. The most important newcomers at this time were two rather similar types, the cycads and the bennettitaleans. These gymnosperms became abundant in Jurassic times and numerous fossils of their large, palm-like leaves have been discovered. The bennettitaleans are now extinct, but several cycads survive in the warmer parts of the world. Most modern forms have a short, columnar trunk, but some Jurassic ones had slender, branching trunks.

Thickets of ferns, including tree ferns, were abundant in Jurassic times, which suggests that the climate was rather

Above: The maidenhair (Ginkgo) tree is a living fossil—the only one of its type still alive. Pollen and seeds are produced on separate trees. This is a male tree, showing the pollen-bearing cones.

Left: Living cycads, showing their thick, columnar trunks and their very large leaves. Most species bear very large cones at the tops of the stems.

warm. Delta deposits contain the fossils of many types of horsetail, similar to those of today except that they are larger. It is believed that these horsetails grew extensively in shallow waters, rather as reeds do today.

The maidenhair tree (*Ginkgo biloba*) flourished over wide areas in Jurassic times, together with many closely related forms. They were all gymnosperms, but their peculiar fan-shaped leaves made them look more like flowering trees. All are now extinct except the maidenhair tree, and this has probably survived only because man has considered it attractive enough to plant in parks and gardens. The conifers were also very important in the Jurassic, and by that time almost all of our present-day genera had appeared.

The Rise of the Flowering Plants

Ferns and gymnosperms continued to dominate world floras in the early part of the Cretaceous Period. One particularly common bennettitalean was *Cycadoidea*, whose spherical trunks have often been found as fossils. But by far the most important fossils in the Lower Cretaceous rocks are the few fruits, leaves, and wood fragments which are believed to have come from flowering plants. These mark the beginning of the final phase in the evolution of land plants. Within 50 million years of their first appearance, land vegetation had been completely trans-

A reconstruction of the Cretaceous bennettitalean Cycadoidea, showing the characteristic leaf scars on the spherical trunk.

PLANTS AND INSECTS

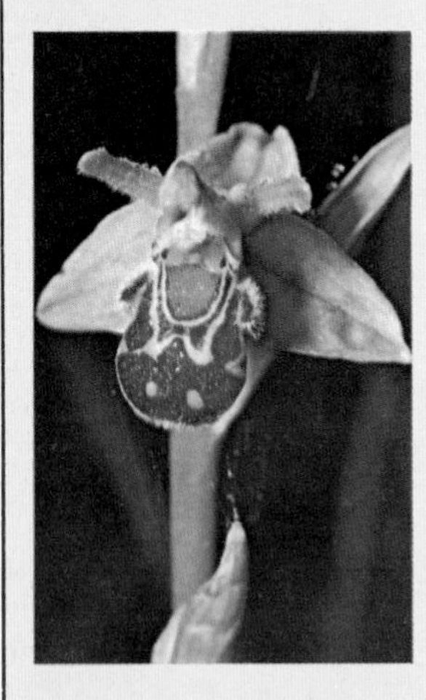

The bee orchid's remarkable resemblance to a female bee attracts male bees, which then pollinate the flower.

The evolution of the flowering plants has been closely linked with that of the insects which rely on them for food. The early flowers were shallow and open to all comers, but later ones gradually became more specialized. Many developed a tubular shape, for example, and concealed their nectar in the tube. The gradual evolution of these tubular flowers was accompanied by the evolution of long tongues in insects such as butterflies, moths and bees. When once such a connexion had been established between a group of flowers and a group of insects, natural selection (see page 27) ensured that the plants and insects evolved together. The flowers which suited the insects best were pollinated the most efficiently and produced the most seeds, while the insects best suited to the flowers got the most food. There are many example of perfect matches between the mouth-parts of insects and the flowers at which they feed, but some of the best examples of co-ordinated evolution are found among the orchids. Some have come to look just like certain female insects, and they are visited and pollinated only by the males of those insects.

formed. By the end of the Cretaceous, the seed ferns and the bennettitaleans had become extinct and the cycads and their maidenhair relatives had become much less common. The flowering plants or angiosperms were rapidly spreading into the spaces left by the vanishing gymnosperms, and they were also spreading out into new habitats.

Flowering plants are a very diverse group, ranging from trees more than 90 metres (300 feet) high right down to the minute, floating duckweeds only a few millimetres across. They also exploit a wide range of habitats. Very few live in the sea, but many have returned to fresh water and, in doing so, have lost many of the features which evolved during the original conquest of the land. Anatomical details, such as the structure of the vascular tissues, help to distinguish flowering plants from gymnosperms, but the essential differences are to be found in their reproductive systems. The flowers themselves vary enormously, from brightly coloured specimens which are pollinated by insects to minute and inconspicuous ones which are pollinated by the wind. But, whether the flowers are brightly coloured or not, the seed-producing structures are all basically the same. The seeds are enclosed in some kind of fruit, and it is this which really separates flowering plants from gymnosperms.

Very little is known about the origin of the angiosperms. It has been suggested that they arose from the bennettitaleans, but this seems unlikely. Study of the early history of the flowering plants is limited in scope because most of the fossils are of leaves. Unless these are found in association with fruits, we cannot be certain that they came from flowering plants. It does seem likely, however, that the earliest flowering plants were evergreen trees and that they arose in tropical areas. They evolved and spread very rapidly in the Cretaceous Period, and there was a further burst of evolution when new mountains appeared in the Tertiary Period and provided new habitats. Climatic changes have since altered the distribution of the various kinds of flowering plants, but today's plants themselves are very little different from those living 60 million years ago.

Magnolias are believed to be the most primitive of the flowering plants. Support for this belief comes from the large, cone-like centre of the flower, on which numerous carpels and stamens are arranged in a spiral manner. The structure of the carpels and the seeds also suggests that the magnolias are more primitive than most other flowering plants.

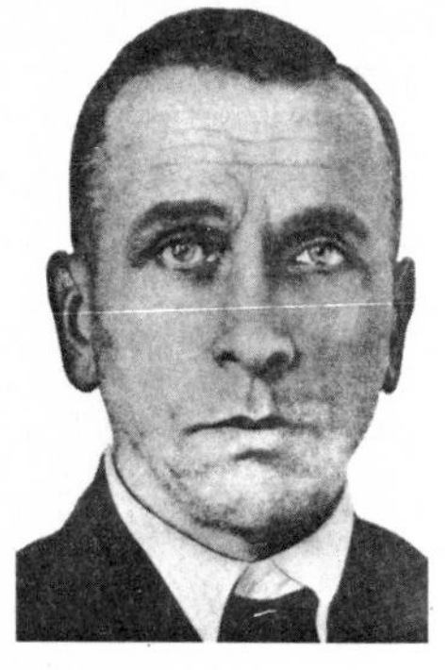

Alfred Wegener

Continents Adrift

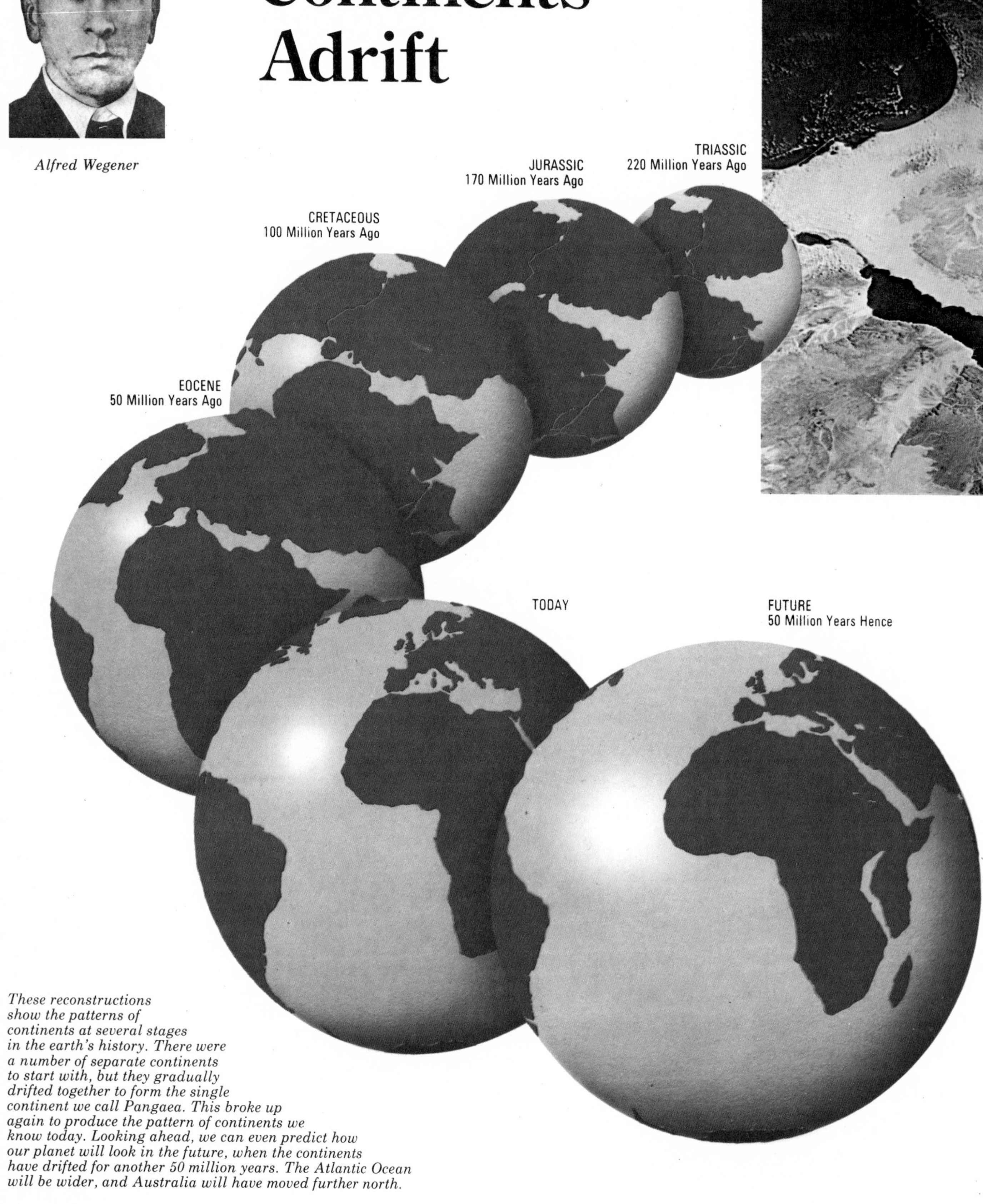

These reconstructions show the patterns of continents at several stages in the earth's history. There were a number of separate continents to start with, but they gradually drifted together to form the single continent we call Pangaea. This broke up again to produce the pattern of continents we know today. Looking ahead, we can even predict how our planet will look in the future, when the continents have drifted for another 50 million years. The Atlantic Ocean will be wider, and Australia will have moved further north.

This satellite picture of the Gulf of Suez and the Sinai Peninsula shows quite clearly how North Africa (bottom left) is separating from Arabia.

As soon as the first crudely-drawn maps of the world appeared, it must have been obvious that the continents would fit snugly together like the pieces of a gigantic jigsaw puzzle. Yet it was not until 1912 that Alfred Wegener (see panel) produced evidence to support the idea that the continents had not always occupied their present positions on the earth's surface, and had probably been connected together in patterns different from those we see today. Geologists rejected the theory of 'continental drift' for a long time because they knew of no force sufficient to split and move continents. But on the basis of more recent evidence the theory has been widely accepted.

Supporting Evidence

The matching outlines of the opposing coasts of the Americas and of Africa and Europe suggested to Wegener (as to others before him) that the continents had somehow split apart. At the time it was not known that the true edge of each continental 'plate' is marked by the continental shelf. This is where the sea bed suddenly plunges from the shallow levels immediately around the continents, to the depths of the oceans that lie between them. Recent computer studies have shown that the shapes of the continental shelves around the Atlantic Ocean can be fitted together very accurately. Australia can similarly be fitted snugly against Antarctica.

Wegener also noticed that, if the continents were fitted together in this way, mountain chains which had formed at similar times in apparently separate continents would be adjacent to one another. These mountains had presumably formed before the continents split and drifted apart. The same was found to be true of individual types of rock, which matched each other when the continents were 're-joined'.

The theory of continental drift explained another of Wegener's discoveries. About 300 million years ago, a great ice sheet had covered large parts of the present southern continents (South America, Africa, Australia and Antarctica), and even peninsular India. The Indian glaciers seemed to have entered that continent from the south – the direction of the present equator. Wegener suggested that all these continents, including India, had then lain around the South Pole. This great southern grouping of continents is called Gondwanaland, while the northern continents (North America and Eurasia) are called Laurasia. Together, they made up the giant continent Pangaea.

The distribution of fossil plants and animals also provided evidence for continental drift. Some 300 million years ago, all the continents of Gondwanaland had a similar flora of early land plants. This was called the *Glossopteris* flora, after the name of one of its characteristic seed ferns (see page 58). Today palaeontologists can show in even greater detail that the patterns of distribution of land animals, at different periods in the past, closely follow the changing patterns of the land as continents split up or joined together. For example, when all the continents were joined together into a single land mass, similar animals were found all over the world. Later, when Gondwanaland and Laurasia again became separate, different faunas were found. Finally, many of the differences between the mammals of today's continents are due to the fact that modern mammals did not appear and start to spread over the earth until after these continents had begun to move apart. Australia, for example, became isolated before the placental mammals could reach it.

ALFRED WEGENER

Alfred Wegener (1880–1930) was an adventurous scientist – at one time, he and his brother Kurt even held a duration-record for ballooning, of 52½ hours. A student of astronomy and meteorology, he also became an arctic explorer, going on expeditions to Greenland in 1906 and 1912 to study the thermodynamics of the upper atmosphere. The idea of continental drift came to him in 1911, but it was not until after his second Greenland expedition that he was able to elaborate his theories in a book, which was published in 1915. Unfortunately his ideas were scorned during his lifetime. His lack of geological training made it easy for professional geologists to dismiss his revolutionary ideas as the superficial theorizing of an amateur.

Wegener returned to Greenland in 1930 to continue his meteorological work on the ice cap, but he died in the November of that year, before he could complete his research.

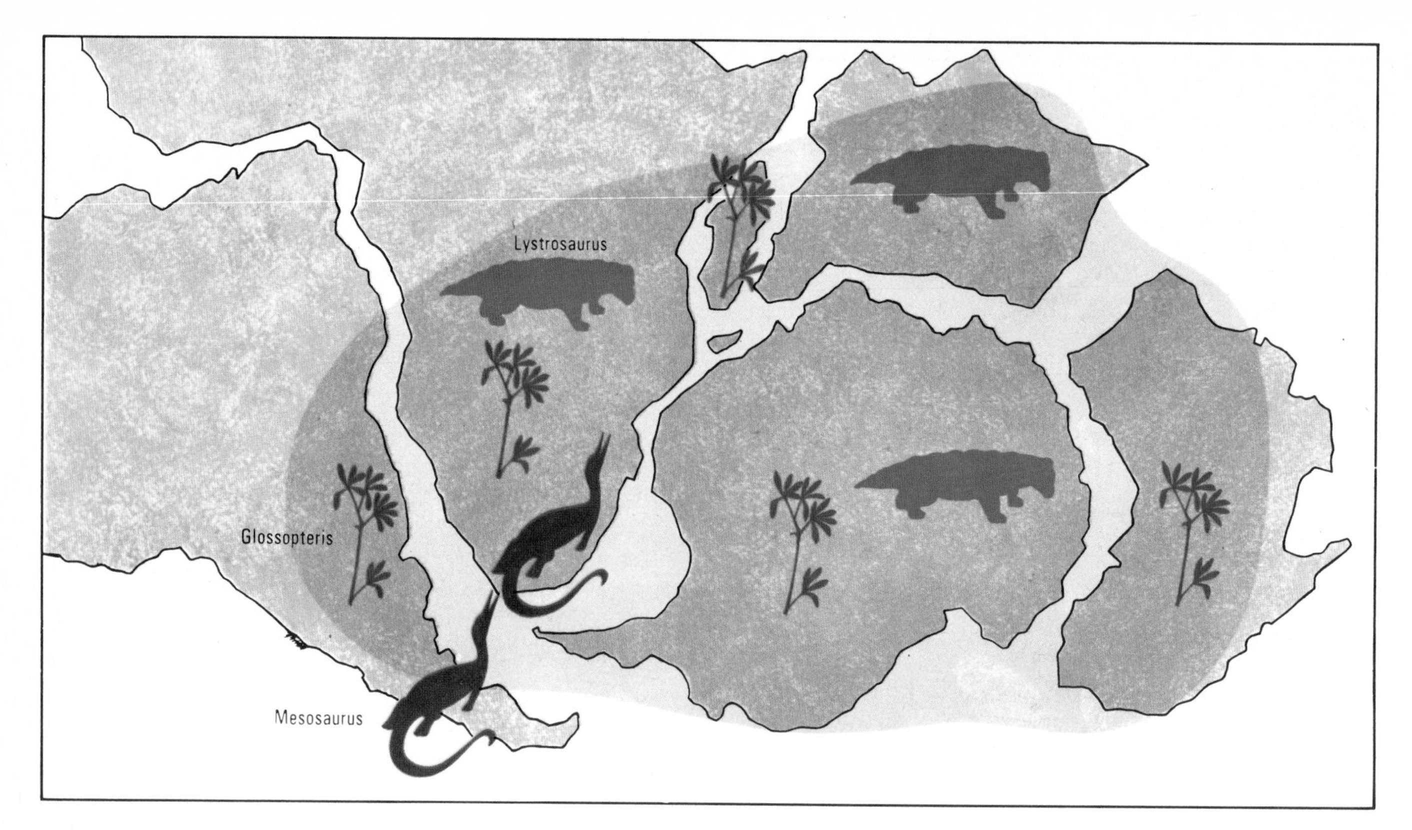

The existence of similar fossils in various parts of the southern hemisphere is strong evidence that the continents were joined together in Permo-Triassic times. The distribution of slightly older, Permo-Carboniferous glacial deposits suggests that the continents had previously surrounded the South Pole. The blue area shows the probable extent of the ice.

The Wandering Poles

More evidence to back Wegener's theory has been supplied by the study of palaeomagnetism (see panel). If we move the continents back along the drift paths indicated by palaeomagnetic studies, we find that they join up in the exact patterns suggested by their outlines and geology. Similarly, the types of rock deposited at different times correspond to their former rather than their present positions. For example, though at first it is surprising to find evidence of glacial activity in the Ordovician and Silurian rocks of north-west Africa, palaeomagnetic data show that the area then lay over the South Pole.

Making Mountains

It is evident from the age and position of various mountain ranges that the continents did not only move apart. Many ranges were crumpled into being as two huge continental masses collided. The Himalayas, for example, are the offspring of a collision between India and Asia. They are comparatively new (only about 25 million years old), and are therefore still high, with peaks over 8,000 metres (25,000 ft), for erosion has not yet been able to wear them down. The Ural Mountains of Russia, which formed when Europe and Asia collided, are much older (250 million years), and erosion has worn them down to less than 2000 metres (6500 ft) high.

The Driving Force

The final, most convincing evidence for continental drift has come, not from the land, but from the floors of the oceans. It is there that

PLUMES

Geologists believe that in some places there are stationary columns, or 'plumes', of heated material ascending from inside the earth, and that where these reach the surface they produce volcanic islands. As the earth's crust moves across such a plume, old islands become disconnected from it, and a new volcanic island forms in the crust above it. With continuous erosion, the old islands grow smaller, eventually disappearing below the surface of the sea.

The Hawaiian chain of islands is a good example. Hawaii itself is the largest and newest – less than one million years old. As one goes westwards, towards the smaller islands, one reaches the older part of the chain, and a series of sunken islands extends right up to the band of North Pacific ocean trenches. Arrows on the diagram below show the directions of movement of the Pacific sea floor at different times in the past.

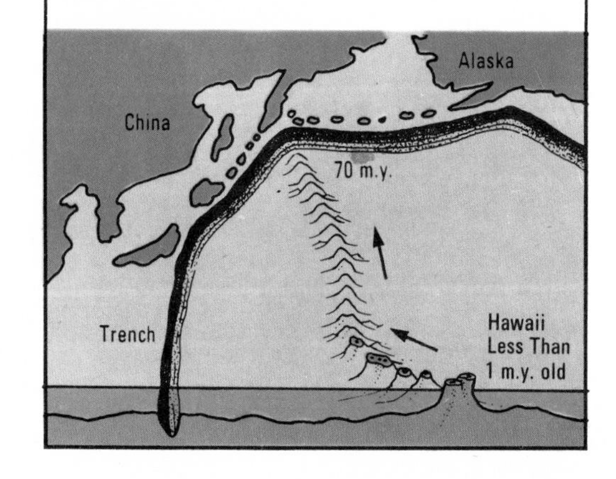

scientists have found the key to the actual mechanism of continental drift – the force that drives continents apart or together.

Oceanographic studies have shown that the earth is girdled by a system of underwater mountain chains, or 'oceanic rises', where much volcanic and earthquake activity takes place. Studies of the sea bed show that the rocks become progressively older as one moves further away from these rises. This is because, as the continents move apart by about 5 to 12 centimetres (2 to 5 inches) each year, new oceanic crust material rises from the deeper layers of the earth and solidifies.

The energy behind this movement comes from within the earth. The deeper layers of the earth are hotter than the surface, and a system of convection currents brings this heat to the surface. There, the currents spread out sideways, pulling the surface crust along with them. This surface crust is really quite thin: compared with the earth as a whole, it is only as thick as the skin on a peach.

Consumed in Fire

If the new sea floor is forming in some places, old sea floor must be disappearing in others – otherwise the earth would have to expand, or its surface buckle. Old sea floor is consumed in a system of great ocean trenches up to 1100 metres (3600 ft) deep. Most of these lie around the edge of the Pacific Ocean. Here lie the return currents of the convection system, recycling cooled material to the deeper layers of the earth. As they descend, these currents pull part of the old ocean bed down with them.

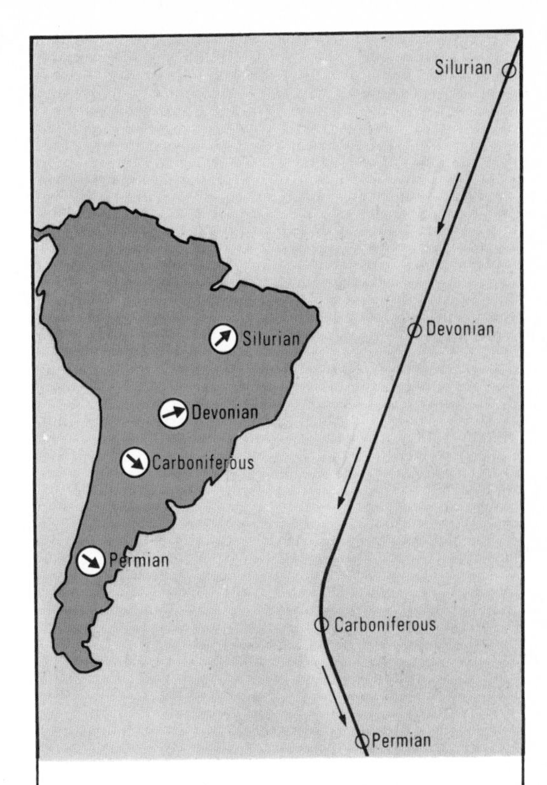

PALAEOMAGNETISM

Some rocks contain magnetized particles of iron-containing minerals which were trapped inside them during their formation. While the rock is still molten, or while it is being deposited, these fossil compass-needles come to point towards the magnetic pole. Once the rock has solidified, this orientation is preserved and can still be detected. By studying such rocks, it is possible to find the direction of the magnetic pole and to calculate its distance.

If the continents had never moved, these tiny fossil compasses would all point towards the present magnetic pole – but they do not. Instead, if a series of rocks of different ages is studied, the compasses point to a series of different positions. These positions show us the apparent movement of the pole past the continent – though really it is the continent that has moved past the pole. The diagram above shows how South America has moved past the southern magnetic pole.

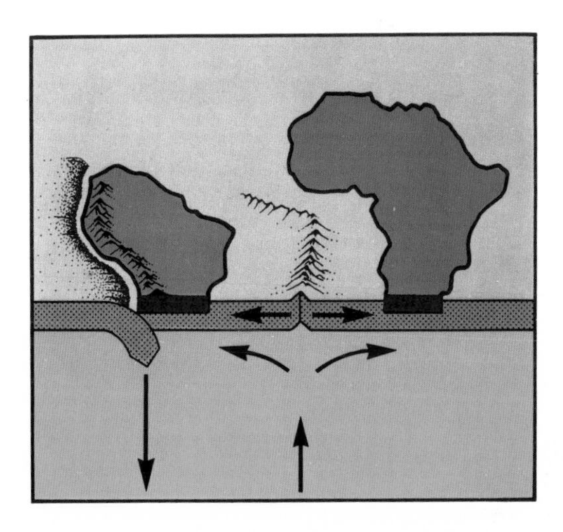

Left: Convection currents within the earth provide the key to continental drift. As the currents spread sideways, thin crustal rocks are dragged with them. The Atlantic Ocean is expanding with the creation of new sea bed at the mid-Atlantic rise. But this expansion is at the expense of the Pacific where old sea floor is disappearing into a trench west of South America.

Below: Volcanoes, mountain chains and earthquakes characterize the separation and collision of crustal plates.

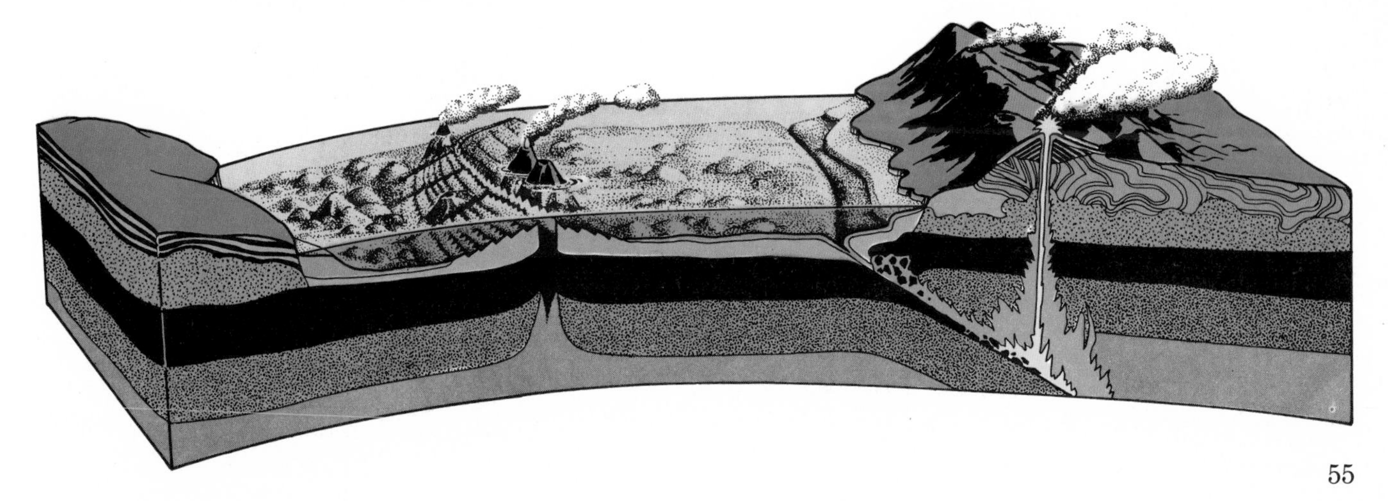

The Palaeozoic World

This towering mass of limestone was formed under the sea during the Permian Period, about 250 million years ago. A study of the minerals and fossils in the rock tells us something of the conditions under which it was formed, while a study of the structural features around the rock may tell us how and when it was pushed up into its present form.

If we could go back to the Palaeozoic Era and circle the earth in a space-ship, we should get a very different view from that seen by today's astronauts. The positions of the continents and the extent to which the sea has covered them have changed many times during the earth's history, and the land masses of 500 million years ago looked very different from today's continents. Some scientists even believe that the earth has expanded since its early days. The Palaeozoic atmosphere probably contained much less free oxygen than the air we breathe today, and there have certainly been important climatic changes. There were, for example, two ice ages during the Palaeozoic Era. It must be remembered, however, that the time available for such changes was immense and that the changes were actually very slow. Similar changes are going on today, but it is difficult to detect them because they are so slow.

Moving Continents

The movements of the Palaeozoic continents are much less well known than those of the later eras, but we can be sure that they did move. Interesting proof of this comes from a study of the Cambrian and Ordovician rocks of Europe and North America. Northern Scotland, western Newfoundland, and eastern Greenland were all receiving the same kind of sediment and it is generally accepted that these regions were close together. They were covered by a shallow sea in which a unique series of sands, muds, and peculiar limestones accumulated. Trilobites are found in the mudstones of Lower Cambrian age and they are identical in all three regions. Northern Scotland, western Newfoundland, and eastern Greenland therefore all belonged to the same *faunal province.* Wales, eastern Newfoundland, and eastern North America belonged to a different faunal province, for the tri-

lobites found in these areas are very different from those of the same age in Scotland and Greenland. The rocks are also very different in the two provinces. Geologists believe that the two provinces were separated by a deep ocean which the trilobites, being shallow-water animals, could not cross. The ocean would have stretched in a north-east/south-west direction across the site of the present Atlantic. Towards the end of the Ordovician Period, however, the two faunas had begun to merge, and by Silurian times the fauna in each region was identical. The explanation is that the two continental masses gradually moved closer together and their native trilobites were eventually able to cross the narrowing ocean. The two continents finally collided and their buckled edges formed a vast chain of mountains.

The Atlantic Ocean has since opened and cut through this ancient mountain chain, but the worn-down remnants of the peaks can still be seen in the Appalachians, in northern Scotland, and in Scandinavia.

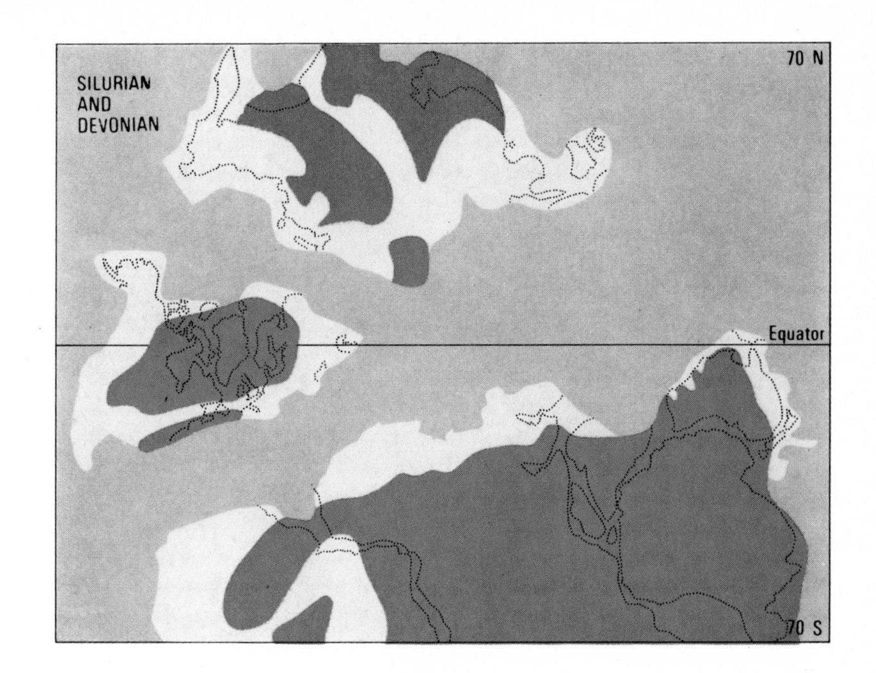

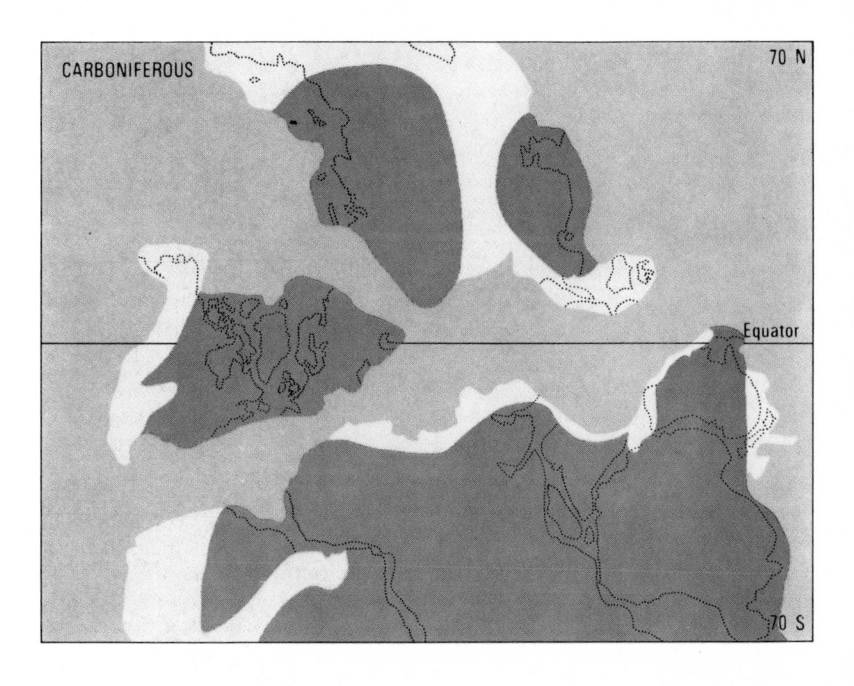

In the Silurian and Devonian Periods (top), there were at least three separate land masses. One was the southern supercontinent called Gondwanaland, another was made up of North America joined to Europe, while Asia formed another. (China may also have been separate.)

In the Carboniferous Period (above), these land masses started to move together and had merged by the middle of the Permian Period, causing great mountain ranges to form – the Appalachian, Allegheny and Ural Mountains.

Palaeozoic Vegetation

The invasion of the land by plants had a marked effect on the earth as a whole. The early land plants added more free oxygen to the atmosphere, stabilized and enriched the developing soils, and altered climatic patterns. They also provided a host of new environments for animals and other plants. By contrast with the rich life of the coal forests, the Cambrian landscape would have been sterile indeed. The rocks may have been encrusted with lichens, and mosses and liverworts may have lived in damp hollows, but we can only guess at this because the earliest-known fossils of these simple plants come from Devonian rocks. If the plants were present in early Palaeozoic times, we can be sure that they had only a patchy distribution. It was only when plants had evolved a vascular system of tubes for carrying food and water that they could colonize the land in large numbers and invade the more difficult environments.

The first fossils of true vascular plants are found in Upper Silurian rocks. The plants were only a few centimetres high, with leafless stems forking into two (see page 44). They bore spore cases on the tips of the stems. Oddly enough, some distant relatives of these primitive plants still live in the tropical forests today.

By the Devonian Period, the early plants had been joined by ferns, giant horsetails, and tree-sized clubmosses, or lycopods. These persisted into the coal forests of the Carboniferous Period, when the early seed ferns and conifers also flourished. The ancestors of the modern maidenhair tree (see page 50) also lived in the later part of the Palaeozoic Era. The southern hemisphere supported an altogether different flora in Carboniferous times. Known as the *Glossopteris* flora, from one of its major components, it was adapted to life in the cooler regions around the South Pole.

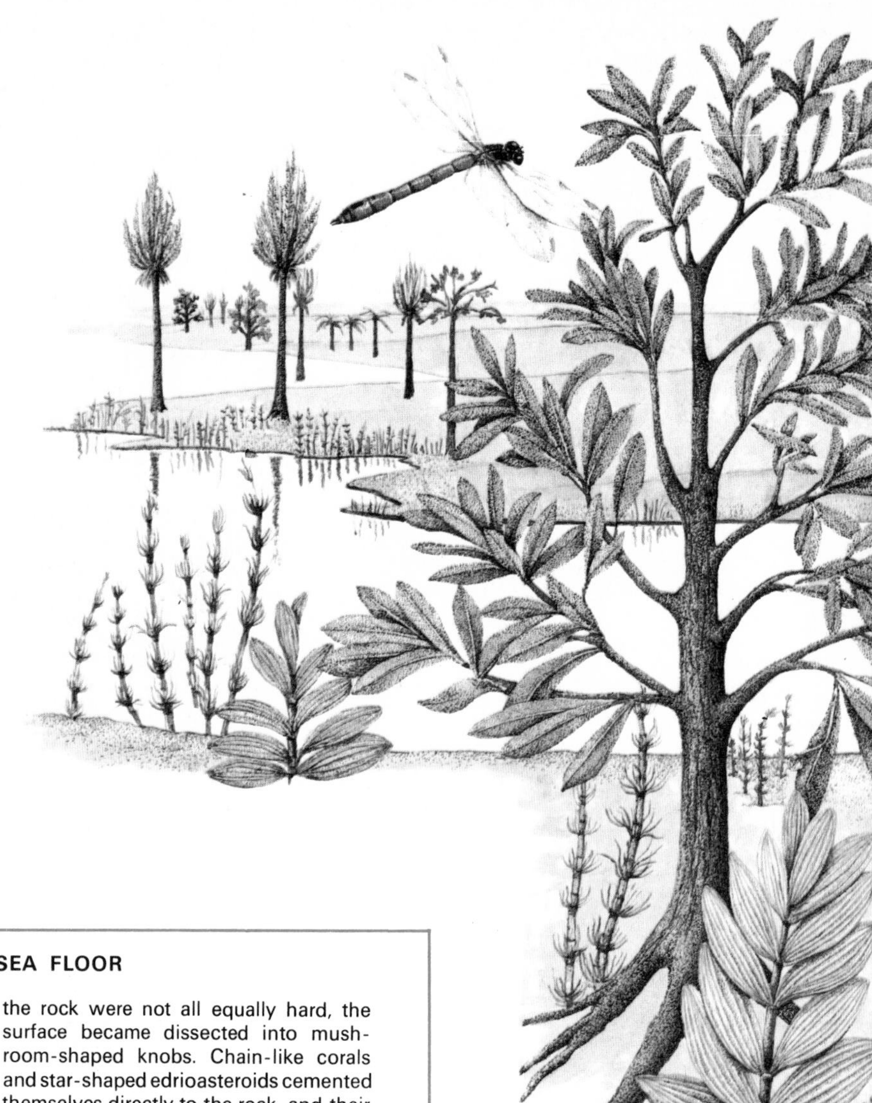

Changing Climates

The rocks indicate that there were general fluctuations in the climate of the Palaeozoic Era. We can trace these climatic changes in some detail for certain areas where there is a good succession of rocks. We can, for example, trace a progression from icy conditions to a warm temperate climate and then to an arid climate before glacial conditions returned. It is more difficult, however, to assess climatic changes on a world-wide scale.

There were certainly widespread polar ice caps just before the beginning of the Cambrian Period, for late Precambrian boulder clays are found in northern Europe, Australia, and several other places. The climate seems to have been much warmer in early Palaeozoic times, even at high altitudes, and the red desert sandstones which covered much of the

A DEVONIAN SEA FLOOR

It is not often that we find in the geological record a piece of undisturbed sea floor with the animals preserved exactly as they lived. What appear to be natural assemblages of fossils have often been brought together by the action of currents. But an example exists from the Devonian of Iowa where there can be no doubt that the animals are in place. An area of hard limestone, probably of Ordovician age, formed a flat pavement on the sea bed. This was eroded by currents, and because different parts of the rock were not all equally hard, the surface became dissected into mushroom-shaped knobs. Chain-like corals and star-shaped edrioasteroids cemented themselves directly to the rock, and their fossils are still found just where they grew. Cystoids, which are extinct echinoderms related to the sea-lilies, also grew there. They are now found lying in the crevices between the 'mushrooms', but the bases of their broken stalks are still to be seen firmly attached to the rocks.

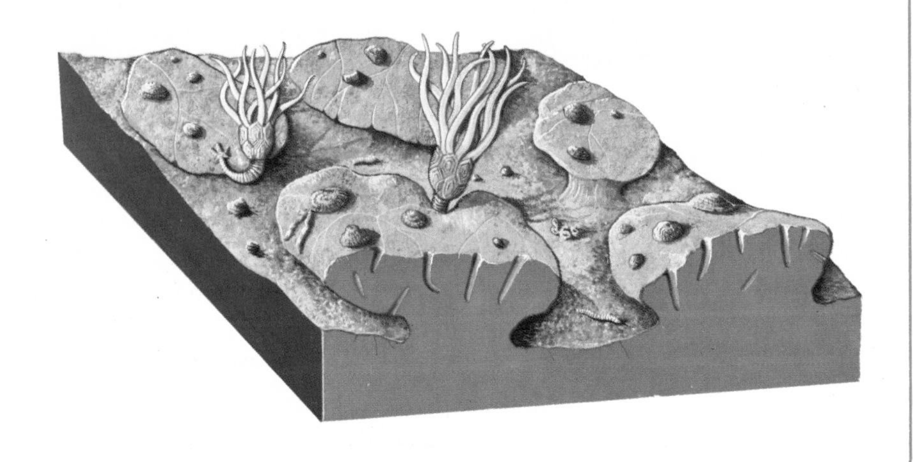

The picture above is a reconstruction of a Gondwanaland scene in late Carboniferous or early Permian times. The scene is dominated by the large-leaved Glossopteris, and other plants include Schizoneura (lower right), Ginkgoites (the fan-shaped tree, middle right), and the horsetail-like Phyllotheca (far left). Fossils of this Glossopteris flora are scattered over a wide area of the southern hemisphere (see page 54), but most of the fossils occur only as scattered leaves and we can often only guess at the appearance of the complete plant. Most of the plants lived in freshwater swamps.

Devonian world show that the climate was sometimes very dry. But there was a glaciation in late Ordovician times in what is now the Sahara region, which was then located close to the South Pole.

Warm conditions prevailed during the Lower Carboniferous and there is no evidence of ice caps at all. By Upper Carboniferous and early Permian times, however, glaciers had advanced over large areas of the southern continents, which were grouped together around the South

Pole and collectively called Gondwanaland. The vastness of the glaciated area shows that this was indeed a very extensive glaciation, yet huge coal forests flourished at the same time in Eurasia and North America, which then lay in the tropics. Much of the Permian Period was very dry, and it was also a time when the seas retreated from the continents and left only shallow, reef-bound lagoons. These lagoons gradually dried up and deposited great quantities of salt which are mined commercially today. There were no glaciations during the Mesozoic Era, and the ice caps may have disappeared altogether for a long period before advancing again in late Tertiary times.

Above: The sloping bedding planes of this sandstone cliff show that it was formed from a desert sand dune.

Below: Examination of the sand grains in a rock may show whether it was formed in water or in desert surroundings. Water-deposited sands contain rather angular grains (top), while desert sands are very rounded.

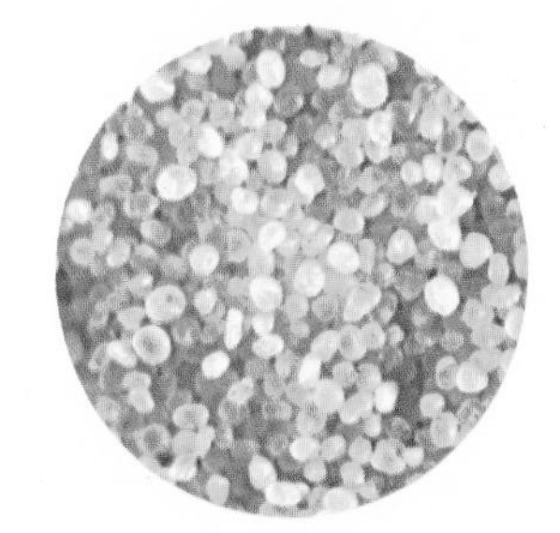

ANCIENT CLIMATES

It is often said that the rocks are like the pages of a history book. They certainly tell us a lot about the earth's history when once we have learnt how to read them. Fossils in the rocks tell us about life in the past, and the rocks themselves may also tell us what the climate was like at a particular period. Beds of boulder clay, accompanied by scratched pebbles and rocks, are clear indications that an area has experienced an ice age or glaciation. Deposits of bauxite, from which we obtain aluminium, provide evidence of tropical monsoon climates. Bauxite is derived from the deep red soil known as laterite, which forms today only in hot lands with strongly seasonal rainfall.

Ancient desert climates can be detected by looking at the sandstones which they left behind. The layers in these sandstones slope first this way and then that, showing how they were built up by winds blowing from different directions. In addition, these desert sandstones are composed largely of neatly-rounded sand grains. Sand grains blowing about on the desert floor are continually bumping into each other and having their corners knocked off, and so they soon become rounded. Seashore and river sands are always more angular, because the water cushions their bumps to some extent. Extensive salt deposits also indicate a very dry climate, for they were usually deposited on the floors of lagoons and shallow seas.

Life in Ancient Seas

The Precambrian rocks contain relatively few fossils, but the beginning of the Cambrian Period is marked by an almost explosive proliferation of highly organized animals. The reasons for this sudden blossoming of the fossil record are now becoming fairly clear. The physical environment was apparently changing quite quickly some 600 million years ago, when the curtain was rising on the Cambrian Period. There was, for instance, a widespread Ice Age during late Precambrian times. Traces of this are found both in northern Europe and in Australia. The cold climate probably augmented the selection pressure on evolving organisms, and so increased the rate of evolution (see page 31). The later warming and flooding of the continental shelves by melted ice may then have given the newly evolved organisms an excellent chance to multiply and diversify even further.

Again, the concentration of oxygen in the air was probably at a very low level until late Precambrian times, and would not have encouraged the evolution of any but the most sluggish animals. Not all scientists accept this idea but, if it is correct, the rapid build-up of oxygen which occurred early in the Cambrian Period would have acted as a spur for the evolution of more mobile animals which use up a lot of oxygen.

Most animal groups evolved very rapidly during the early Cambrian. One of the most important changes was that several kinds of organism independently acquired the ability to make hard shells for themselves. This was a big biochemical advance, which may well have been encouraged by the fact that, at this time, there was more lime available for making shells. The lime resulted from the gradual clearing of atmospheric carbon dioxide by algae in late Precambrian times. Some of the carbon dioxide was incorporated into calcium carbonate by the lime-secreting algae, and some of the lime deposited by them was then dissolved in the water.

Exquisitely preserved in the Burgess Shales, this fossil of the shrimp-like Waptia clearly shows the head shield, the antennae, and the feathery limbs which were used for swimming.

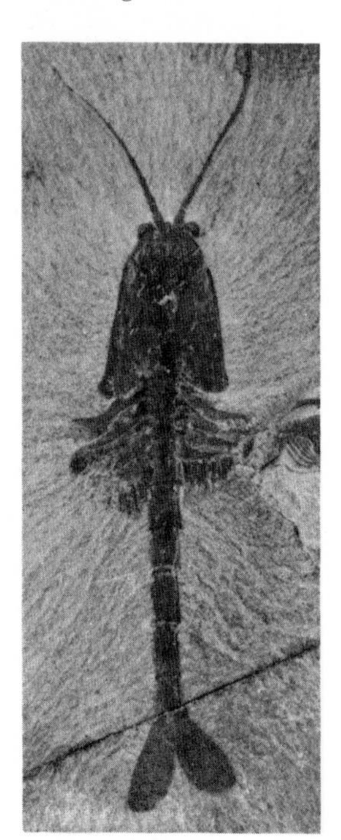

A WINDOW ON THE CAMBRIAN

Cambrian fossils are very widely distributed and occur in many kinds of rock, but there is one rock formation which emphasizes above all others the amazing diversity of life in Cambrian times. This rock is the Burgess Shale, which dates from the middle Cambrian and which is known only from two small quarries high up in the mountains of British Columbia. Its extraordinary assemblage of fossil animals was discovered around the beginning of the 20th century by an American geologist named Charles Walcott. The animals are preserved as thin, delicate films in a very fine, dark shale. It seems that these little creatures lived at the base of a submarine cliff and that a river of mud poured over the cliff at some time, trapping the animals and preserving them for ever in the exceptionally fine sediment.

The shale contains many trilobites, some of them with their limbs intact. But these look very large and clumsy beside the delicate crustacean-like creatures which accompany them. The latter include the remarkable *Marrella*, with four horns springing from its head, and *Leanchoilia*, which has long pincer-like claws on each side. There is also the shrimp-like *Waptia*, and the strange *Burgessia*, under whose umbrella-like shell can be seen the feathery branches of the gut. Worms, sponges and jellyfish are present in large numbers, and there is also the caterpillar-like animal called *Aysheia*, which resembles the modern velvet worm or *Peripatus*. But *Peripatus* lives in the tropical forests and *Aysheia* was a marine creature.

Were it not for the Burgess Shales, laid down in rather freak conditions, palaeontologists might have a very different view of Cambrian life. It is only through the 'window' of the Shales that the real diversity of this ancient fauna can be seen.

As soon as the marine animals were able to make hard shells, they could be preserved as fossils. And so, from the beginning of the Cambrian, we suddenly have a clear record of all kinds of animals: trilobites, brachiopods and other descendants of late Precambrian creatures whose organic shells were too thin and fragile to have been fossilized.

Giants of the Seas

The trilobites were all marine animals. Their fossils first appear in Lower Cambrian rocks. A great many species evolved during the Cambrian and Ordovician times, and their fossils are very useful for zoning and dating the rocks of these periods. After the Ordovician Period, however, the trilobites began to decline; the last survivors became extinct in Permian times.

Most of the trilobites were from two to ten centimetres (four-fifths of an inch to four inches) long, although there were some giants of about 70 centimetres (28 inches) and many dwarfs only three millimetres (one-tenth of an inch) long. Like all other arthropods, including the insects and crabs, the trilobites had a hard outer shell and a number of jointed limbs. The shell, or cuticle, was made largely of a form of limestone called calcite. The head was a rigid plate, while the thorax consisted of a number of segments rather like those of a woodlouse's body. The tail region was another flat or curved plate with transverse divisions. The central part of each region was raised up above the rest of the body, thus dividing the body

Above: Many trilobites were blind, but others had very efficient compound eyes made up of numerous separate lenses, as can be seen in this highly-magnified photograph. Their eyes were thus similar to those of insects.

Below: Judged by the number and variety of fossils found in the Cambrian rocks, the Cambrian seas must have teemed with life. The shapes of some of the fossils together with the positions in which they were fossilized, help us to work out how the animals might have lived. The picture below shows something of life in early, middle and late Cambrian times. Although some creatures were confined to one or other of these divisions, many of them survived throughout the Cambrian Period.

KEY

1. Several kinds of sponges; **2**. Jellyfishes; **3**. *Olenellus* (a trilobite); **4**. Red algae; **5**. *Ellipsocephalus* (trilobites); **6**. *Aysheia*; **7**. *Burgessia*; **8**. *Nisusia* (brachiopods); **9**. *Emeraldella* (a crustacean); **10**. Corals; **11**. *Waptia*; **12**. *Marrella*; **13**. Green algae; **14**. *Paradoxides* (a trilobite); **15**. *Cruziana* (a burrowing trilobite); **16**. *Billingsella* (brachiopods); **17**. Crinoids; **18**. Echinoderms; **19**. Ragworm

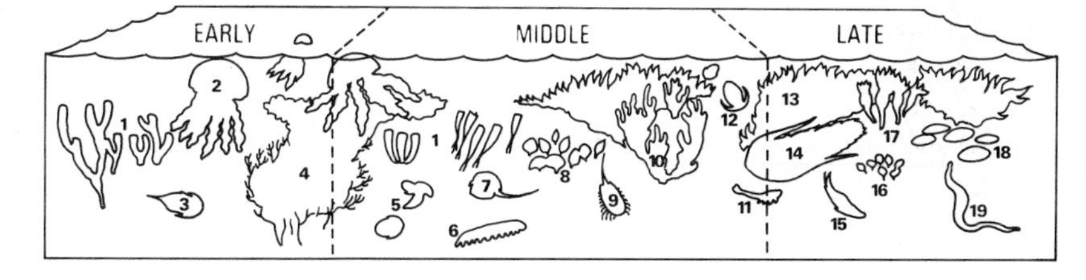

Above: A giant among trilobites, this specimen of Paradoxides was about 30 centimetres (one foot) long. The fossil was preserved in Middle Cambrian rocks in Newfoundland.

Left: A series of graptolites showing how the branching became less complex as the animals evolved. From the top, the fossils are: Dictyonema (Cambrian), Dichograptus, Tetragraptus, and Dicranograptus (all Ordovician), Rastrites and Monograptus (both Silurian).

into three longitudinal lobes which are responsible for the name 'trilobite'. Most trilobite fossils are incomplete, and very often only the head or tail regions are found, but specimens complete with legs have occasionally been discovered in the finer sediments. The complexity of their structure shows that the trilobites were certainly not low-grade or inefficient animals; though primitive in some respects, they were highly evolved and very successful in their own environment. In times of danger, some trilobites were able to roll up, rather as a woodlouse does.

Some of the smaller species may have floated freely in the plankton, but most of them probably crawled on the sea bed or swam slowly over it by means of their feathery limbs. Some of the more spiny kinds were probably able to rest on the sea bed on the tips of their spines. The chance fossilization of tracks in the mud has shown how some of the trilobites moved about. Some walked on the tips of their legs, and curious little radiating marks show where the claws on the ends of the legs were placed one after the other on the mud. Such tracks show that some trilobites were even able to walk sideways, pulling themselves crabwise over the sea bed and leaving little scratch marks in the mud. Deeper trails were made by trilobites ploughing their way through the mud. Such trails take the form of paired grooves with chevron-like furrows crossing them. The trilobites that made them were probably scavenging for food in the mud, picking up material from the sea bed or else filtering food particles from the water.

Graptolite Graffiti

Like cryptic messages scrawled across the rocks, the fossil traces of graptolites provide a fascinating code for palaeontologists to decipher. The remains of these ancient animals are usually found in dark shales and, on close inspection, they resemble little saw-blades. They occur only in Palaeozoic rocks, and are most abundant in Ordovician and Silurian strata. Because of their wide distribution and rapid evolutionary changes they have proved very valuable in zoning and dating great thicknesses of Lower Palaeozoic shales which are otherwise devoid of fossils (see page 33).

The earliest graptolites (called dendroids) appeared in late Cambrian times. They had many branches, sometimes joined together by narrow tubes. Each of the branches was lined with tiny cups called *thecae*, and it is these cups which give the fossils their toothed appearance. The thecae apparently held the *zooids*, which were the living parts of the graptolites. Each zooid had its own cup, but it was linked to all the other zooids on the branch by slender canals. The graptolites are thus the earliest known colonial animals.

The dendroids had two types of thecae, but later graptolites had only one type. The whole story of graptolite evolution has, in fact, been one of continued simplification. We can see a gradual progression from the many-branched dendroids of the late Cambrian, through early Ordovician forms with eight or four arms, to Upper Ordovician graptolites with just two arms or branches, and finally to Silurian species with only a single spike. The names of the various genera often reflect the number of branches present: thus we have *Tetragraptus* with four branches, *Diplograptus* with two, and *Monograptus* with just one. The shapes of the branches and thecae also vary, and it is this complex array of different species that makes the graptolites so useful in stratigraphy – the science of determining the chronological order of rock layers. Each kind of graptolite lived only for a limited period of time, and the palaeontologist finding *Monograptus* fossils in a rock sample will immediately know that the rock is of Silurian age. What is more, by identifying the exact species of *Monograptus*, he can say whether the rock was laid down in early, middle, or late Silurian times.

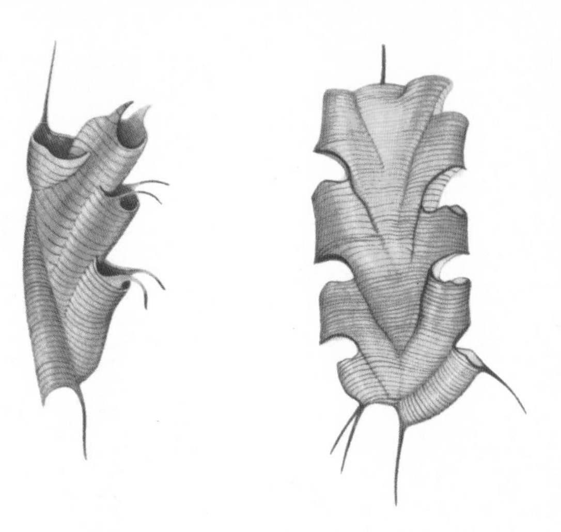

Right: Drawings showing the detailed structure of the graptolite cups, or thecae. Monograptus, on the left, has cups only on one side of the stalk, while Amplexograptus has them on both sides. The lowest cups were formed first, and further cups were added as the colonies grew and the stalks became longer. The fine lines across the cups represent the daily growth lines, but we still know almost nothing of the animals that lived in these cups.

Graptolite fossils are sometimes found in which the original horny skeleton has remained almost unaltered and uncrushed. The zooids have disappeared, of course, but we can see the wonderful complexity of the skeleton, even down to the minute daily growth lines. Such detail is quite impossible to detect in the usual flattened fossils, and palaeontologists did not even know what group the animals belonged to until these three-dimensional fossils had been found. The graptolites were once thought to be relatives of the corals, but they are now known to be related to the pterobranchs – rare animals which live in today's oceans and belong to the same major group as the sea squirts and the vertebrates. Even now we do not know what the little graptolite zooids looked like. We can only guess that they were tiny coral-like creatures that filtered food from the water with their tentacles.

MONSTER ARACHNIDS

The strangest and largest arthropods ever known were the eurypterids, or sea scorpions, which thrived from Ordovician times until the Carboniferous Period. They were very like modern scorpions in appearance, with a distinct head, a swollen and segmented body, and a tapering tail region. Six pairs of appendages were attached to the underside of the head. In the earlier eurypterids of Ordovician and Silurian times, such as the genus *Eurypterus*, the first pair of appendages were little jaws. These were followed by four pairs of walking legs, and a curious pair of paddle-like legs which were perhaps used mainly for swimming.

Most of the early eurypterids were less than ten centimetres (four inches) long, and they probably all lived in coastal waters and brackish estuaries. By late Silurian times some enormous species with a truly terrifying appearance had emerged. *Pterygotus* was two metres (six feet) long, with huge predatory pincers. One of the largest invertebrates which have ever lived, it probably ate fishes and other eurypterids. *Carcinosoma* was a fat-bodied form with a poison spine on its tail, while *Stylonurus* crept around the sea bed on long, thin legs. Some eurypterids had large eyes with many lenses set at the side of the head; others had small eyes in the middle of the head.

By the Carboniferous Period most of the huge forms had died out. Their place was taken by smaller species which, for the first time, were colonizing truly freshwater environments. Then they all became extinct. Although the eurypterids looked quite like true scorpions, the two groups were not closely related and the eurypterids did not give rise to modern scorpions. The early scorpions were marine animals, and we do not know when their descendants made the difficult transition from the water to the land.

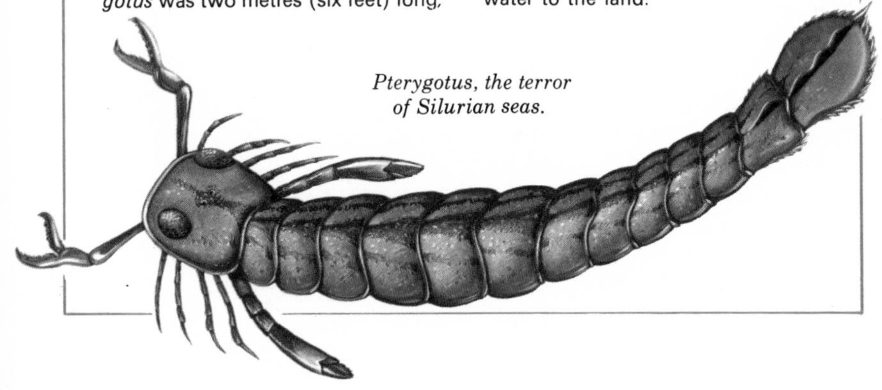

Pterygotus, the terror of Silurian seas.

A crinoid, or sea-lily, and two small coral colonies fossilized on a slab of limestone.

The early dendroids may have been fixed to the sea bed, but all the later graptolites were planktonic, floating in the upper layers of the sea and possibly suspended from floating seaweeds. When they died, they sank to the sea bed. Graptolites were normally preserved only in the finer sediments, which later became shales and limestones. Their delicate structure would not have survived strong sea-bed currents and they are hardly ever found in sandy sediments.

Ancient Echinoderms

Starfish and sea urchins are familiar animals which show very clearly the characteristics common to all echinoderms. They are all marine creatures with an internal skeleton composed of calcium carbonate plates embedded in or just under the skin. They usually have a five-rayed symmetry, which is not found in any other group of animals, and they have a peculiar internal system of pipes and bladders called the water vascular system. It is this system which enables the echinoderms to move, feed and breathe. In a sea urchin, for example, the system consists of five main water-filled tubes from which spring a number of bladders. The latter lead to

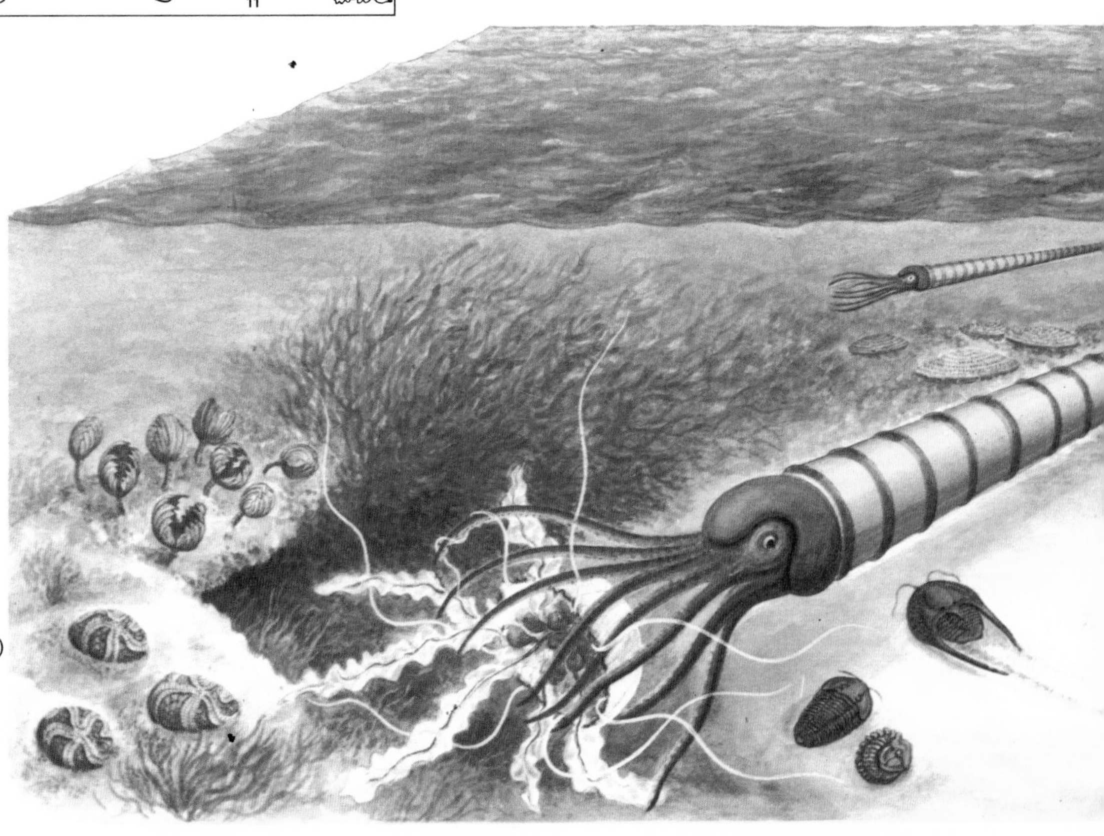

Some of the creatures which lived in the Ordovician seas.

KEY

1. Edrioasteroids (early echinoderms);
2. Brachiopods;
3. Algae (seaweeds);
4. Coral colonies
5. A large nautiloid capturing a jellyfish with its 'arms';
6. *Calymene* (trilobites) one of which has rolled up like a woodlouse;
7. *Cryptolithus* (a trilobite) crawling over the mud and leaving its characteristic chevron-like trails;
8. *Isoteloides* (a trilobite) crawling crab-fashion across the mud;
9. *Placocystis* (an echinoderm) anchored by its 'tail' but lying flat on the sea bed;
10. An underwater 'forest' of bryozoans, with a cystoid clinging to one of them;
11. A group of *Dinorthis* brachiopods.

sucker-like tube-feet which pass out through the body wall. When the bladders contract, water is forced out into the tube-feet and these extend to several times their original length. The extended tube-feet can grip rocks and seaweeds with their suckers and then, by contracting and forcing the water back into the bladders, they can drag the animal along. Starfish move in the same way.

The fossil record of the echinoderms is very rich because their calcite plates are resistant to decay and easily fossilized. Many limestones contain a high proportion of echinoderm remains, and these deposits show that the echinoderms were much more numerous and more varied in the past than they are today. The earliest echinoderms lived in the Cambrian Period, but by Ordovician times they had reached a complexity which in some instances seems far greater than that of many modern forms. Some of these ancient creatures were so bizarre that, in spite of their calcite plates, it is difficult to be sure that they were echinoderms at all.

Fossilized sea urchins (Hemicidaris). Their long spines were attached to little knobs on the shell, forming ball-and-socket joints, and thus enabling the animals to move about on the sea bed.

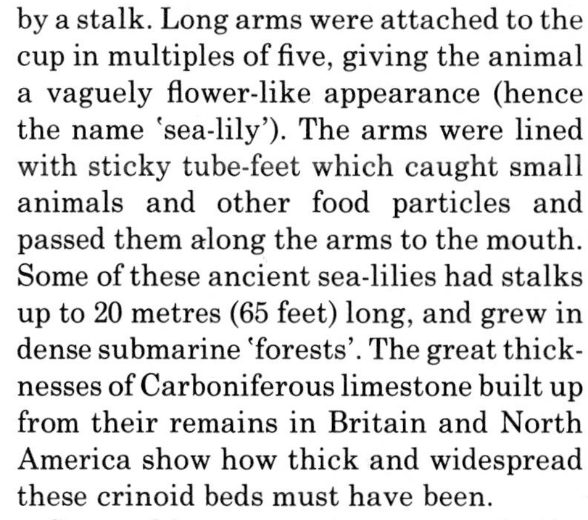

The sea-lilies, or *crinoids*, and their extinct relatives were very common in the Palaeozoic Era. They had a fairly small, cup-shaped body attached to the sea bed by a stalk. Long arms were attached to the cup in multiples of five, giving the animal a vaguely flower-like appearance (hence the name 'sea-lily'). The arms were lined with sticky tube-feet which caught small animals and other food particles and passed them along the arms to the mouth. Some of these ancient sea-lilies had stalks up to 20 metres (65 feet) long, and grew in dense submarine 'forests'. The great thicknesses of Carboniferous limestone built up from their remains in Britain and North America show how thick and widespread these crinoid beds must have been.

Sea urchins were not numerous in the Palaeozoic Era, and those which did exist were quite simple in form. They lived in quiet lagoons, feeding on algae and debris on the sea bed. Although some species were quite large, they were not a very vigorous group of animals. Starfish were well established early in Palaeozoic times, and apparently behaved much as they do now. Today's hungry starfish settles over a cockle or some other bivalve, fixes its tube-feet on to the shell, and uses their powerful suction to pull the valves apart. The association of starfish fossils with those of dislocated and broken bivalve shells suggests that this feeding method was in use as far back as Devonian times.

Among the many other groups of echinoderms living in Palaeozoic times were the brittle stars, the sea cucumbers, the blastoids, and the cystoids. The last two groups were relatives of the crinoids.

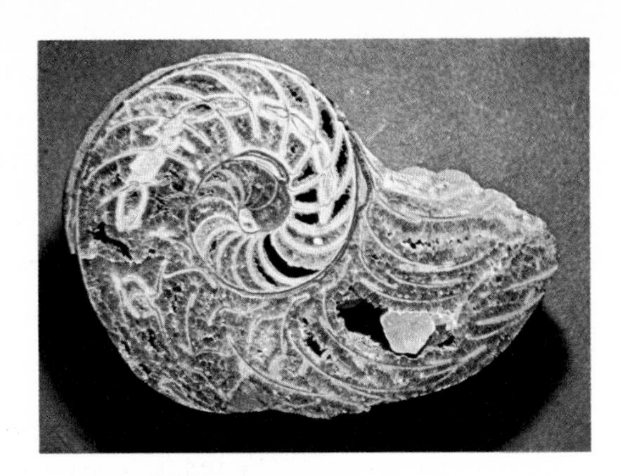

Right: A fossilized ammonite cut through to show the numerous chambers and partitions.

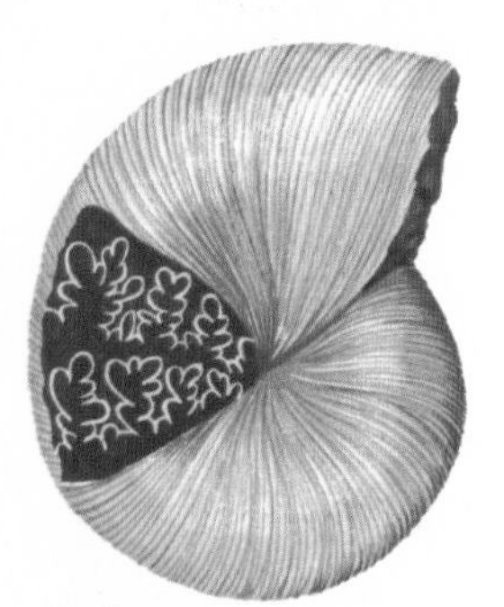

Phylloceras (with part of the shell removed to show the very complex suture lines)

Ceratites (with wavy, but still fairly simple sutures)

Goniatites (with zig-zag sutures)

Extinction Escaped

Some of the echinoderms were short-lived evolutionary experiments, but others had a longer history and some flourished throughout the Palaeozoic Era. Most of the ancient echinoderm groups reached the end of the line in the Permian Period, however, when widespread extinction affected nearly all groups of marine animals.

The starfish and brittle stars survived the Permian extinctions quite well, but the crinoids and sea urchins fared less well. All the Palaeozoic crinoid groups died out, but one versatile new group was already in the making, and it became well-established in the Triassic Period. All later crinoids were descended from this one group.

Only one genus of sea urchins survived the Permian extinctions, but as with the crinoids this was the ancestor of a very versatile stock. Its descendants underwent a tremendous burst of adaptive radiation in the Mesozoic Era, and ventured into almost all marine habitats.

Shells of all Shapes

The molluscs are a very large group of invertebrate animals. They include the slugs and snails, the bivalves, the squids and octopuses, and several other smaller and less common groups. There are also several extinct groups of molluscs. The animals are very varied in appearance, but they nearly always have a shell made from some form of limestone.

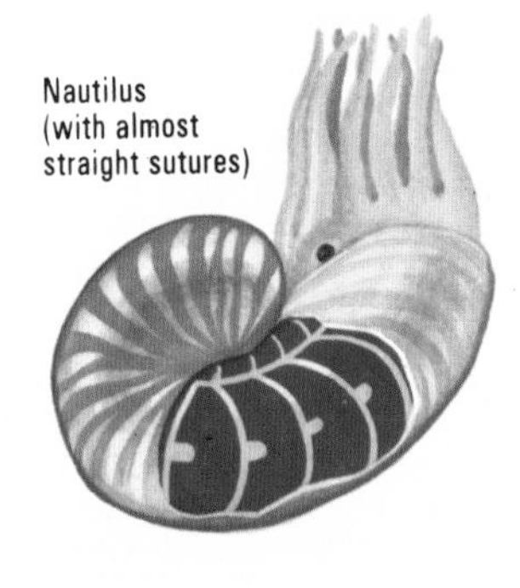

Nautilus (with almost straight sutures)

The ammonites evolved from nautiloids by way of the goniatites, and there was a progressive increase in the complexity of the suture lines – the junctions where the internal partitions met the outer shell. The complex sutures of the ammonites may have made their shells much stronger, and this may be why they gradually ousted the nautiloids and goniatites.

Molluscs are very well represented in the fossil record, the best-known fossils being the ammonites which were so common in Mesozoic seas. Ammonites belong to the same group of molluscs as the squids and octopuses, a group known as the cephalopods. Their nearest living relative is the pearly nautilus, and it would probably be very difficult to understand the ammonite fossils if we did not have the nautilus for comparison. The nautilus has a coiled shell which is divided into several chambers. Most of these chambers are filled with gas, while the animal itself occupies the last and largest chamber. The animal is rather like an octopus: it has tentacles lined with suckers and a powerful beak-like mouth. When the body grows too large for its chamber, it secretes another one and moves forward into it, closing off the last one with a new partition. But the body remains in contact with all its old chambers by means of a fleshy tube called the *siphuncle*, which passes back through all the partitions. The nautilus can swim rapidly by jet propulsion, sucking water into the mantle cavity and squirting it out through a funnel. The direction of the jet can be altered, and the animal can also control its buoyancy. The extinct ammonites and their relatives were probably quite similar in anatomy to the pearly nautilus, but their soft parts are not usually preserved, and almost all our knowledge of them is based on their fossilized shells.

The earliest known cephalopods, the ancestors of the nautilus and of the ammonites, date from late Cambrian times. Their shells were straight or only slightly curved. These were the animals known as nautiloids. Some straight-shelled species from the Ordovician rocks show coloured zig-zag markings which may have been a form of camouflage, for these animals were probably bottom-dwellers, incapable of rising far above the sea bed. Spirally coiled nautiloids did not become common until the Devonian Period. A new group of coiled cephalopods also came into existence at this time; these were the ammonoids, the fore-runners of the am-

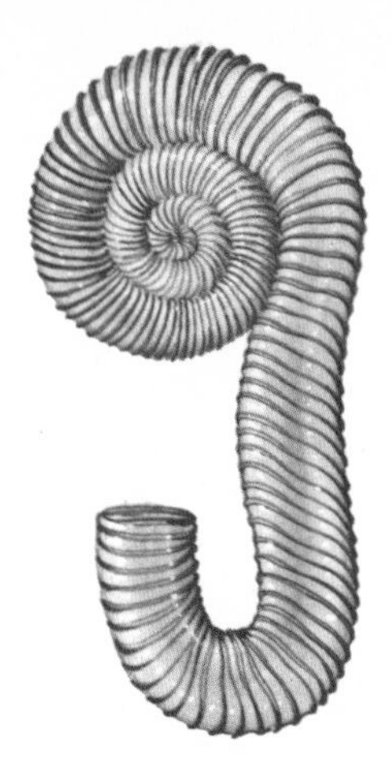

Macroscaphites, a Cretaceous ammonite whose shell is partially uncoiled.

monites which became so common in the Jurassic and Cretaceous Periods.

The ammonoids obviously arose from some kind of nautiloid, but they had a number of important distinguishing features. The earliest ammonoids, called goniatites, were abundant in late Palaeozoic times, but by the Triassic Period they had given rise to the true ammonites with their very complex structures. For some unknown reason, the ammonites nearly became extinct at the end of the Triassic Period, but they picked up again and underwent a tremendous radiation in Jurassic and Cretaceous times. Different species, identified by differences in their shell structures, followed one another fairly quickly and their fossils are therefore very useful for zoning and dating the rocks. Being mainly floating animals, the ammonites spread throughout the world's seas, and this makes them even more useful because the same species turn up in Australian rocks as in European deposits. Towards the end of the Cretaceous Period, however, the ammonites began to decline. A number of very unusual forms were produced at about this time, including some partially uncoiled species and some which coiled up like snail shells. These experimental forms failed to save the ammonites, however, and the group finally became extinct about 65 million years ago.

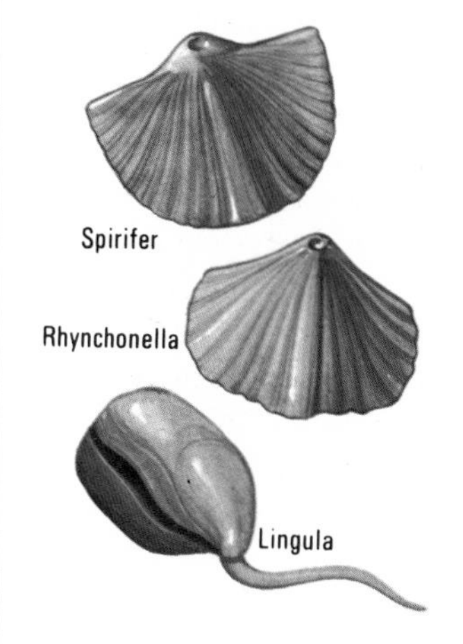

Above: Brachiopod shells can be distinguished from bivalve molluscs because each shell is always symmetrical.

Below: A fossil graveyard, composed mainly of brachiopods.

BRACHIOPODS

The brachiopods, or lamp shells, are among the commonest of all fossils, being found in rocks of all ages from the earliest Cambrian to the Upper Tertiary. There are two or three hundred species living in the seas today, but these are only a small remnant of the 30,000 species known to have lived in the past. By far the greatest proportion of these extinct species lived in the Palaeozoic Era.

Like bivalve molluscs, brachiopods feed by drawing water into the two-valved shell and straining out food particles. The earliest brachiopods had horny shells, held together in a rather makeshift way. Some of these brachiopod species are still alive today, the best known being a type called *Lingula*. This remarkable creature has survived for 500 million years with hardly any change; it certainly deserves to be called a 'living fossil'. *Lingula* is one of the few brachiopods which live in burrows. It digs itself into the sea bed with a long, spoon-like stalk called a pedicle. *Lingula* prefers brackish water, an environment which most other organisms have found difficult to colonize. It has therefore had relatively little competition from other organisms, and this may be why it has changed so little during its long history.

More numerous in the fossil record are the calcareous-shelled brachiopods, whose chalky valves are held together with teeth and sockets. Some of these 'hinged' brachiopods had long stalks which attached them to rocks and held them above the sea bed, but many others lay flat on the sand or mud. Each kind had its preferred habitat and many species were fussy about the depth of water in which they lived. Some Silurian brachiopod communities are now found fossilized in bands running parallel to the ancient shore-lines, each band corresponding to a certain depth of water.

Brachiopods are abundant in Silurian rocks, and even more so in the limestones of the Carboniferous Period. *Gigantoproductus*, found in the Carboniferous limestone, reached diameters of more than 50 centimetres (20 inches), though the majority of brachiopods were smaller. 'Experimental' forms proliferated; by Permian times, a number of species had developed which were so highly specialized that they hardly looked like brachiopods at all. But they failed to survive the widespread extinctions which occurred during this period. Only two groups succeeded in doing so, and although brachiopods became abundant again in the Jurassic, they never regained the diversity and dominance that they had enjoyed during the Palaeozoic Era. In the race to colonize the shelves of the Mesozoic seas, the brachiopods were beaten by the more versatile bivalve molluscs. Living brachiopods are still plentiful in the shallow waters of the Indo-Pacific region, and there are some around the British coasts; but they are vastly outnumbered by the bivalves, and it is probable that they will eventually become extinct in the face of this competition.

The Belemnites

The bullet-shaped belemnites are very well-known fossils. They are actually the fossilized skeletons of squids which lived in the Jurassic and Cretaceous seas. The main part of the belemnite is the *guard*, which is a spike of calcite. Complete specimens have a conical cavity in the blunt end, into which there was inserted a delicate chambered shell. The whole skeleton was housed inside the squid's body, giving the animal a rigid attachment for its muscles. The little chambered shell helped to give it buoyancy. Specimens have occasionally been found lying flat on the bedding planes of the rocks, with the 'bullet' surrounded by a carbon film of the body. Such fossils show the impressions of eight arms with the suckers and hooks in place. The animals obviously fed on fish and other sea creatures just like modern squids, and it is known that they were in turn eaten by ichthyosaurs.

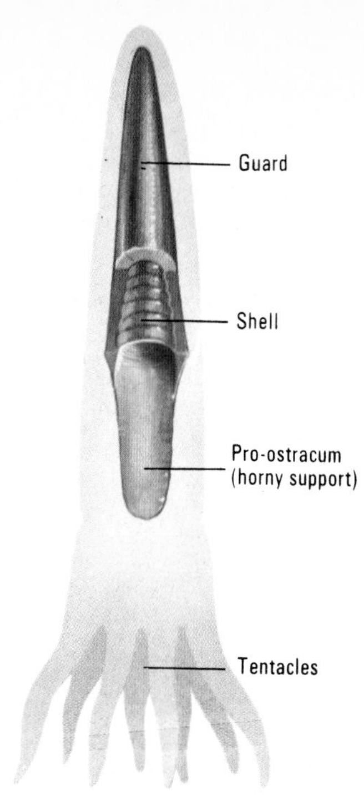

Above: A diagram of a belemnite, with the guard cut away to show the shell. The two longest tentacles are not shown.

Above: Tryplasma, one of the horn-shaped corals which must have lain on the Silurian sea bed.

Snails and bivalves have existed in the seas since Cambrian times, and are now among the most abundant animals of the sea bed. But they did not really become dominant until the Tertiary. The extraordinary Pliocene and Pleistocene crags (see page 20) consist almost entirely of the remains of these animals.

Below: A drawing of an archaeocyathid, an important Cambrian reef-former.

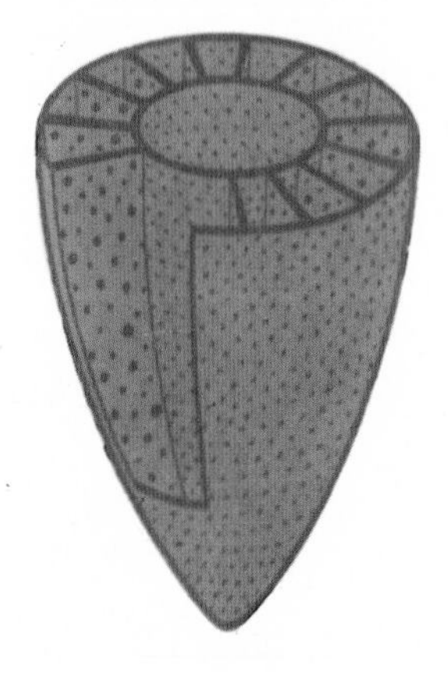

The Reef-Builders

The corals are allied to the jellyfish and the sea anemones, and belong to a group known as the coelenterates. All of these animals have a soft body with a single opening and a ring of tentacles. Although of rather simple construction, the coelenterates are voracious predators. Their tentacles are armed with powerful stinging cells which act like tiny harpoons to catch and kill passing fishes and other animals.

The corals are coelenterates which surround their bodies (or polyps) with limestone cups. Some are solitary creatures, but many species form large colonies in which numerous polyps live side by side. These are the reef-building corals of the warmer oceans. The cups or skeletons of dead polyps form huge masses of limestone and the living polyps form a thin 'cloth' on the surface.

The limestone cups of the corals fossilize very well, and we can follow their record right back to the Ordovician Period. Although many different kinds of corals lived during Palaeozoic times, none of them ever really mastered the art of fixing themselves securely to the ground. Many colonies had pointed bases and, unable to attach themselves, must have formed shifting beds on the sea floor. Their inability to secure themselves meant that the Palaeozoic corals could not form large reefs. The extensive reefs during this period were mainly formed by other organisms. Like many other groups of marine animals, the corals declined suddenly at the end of the Palaeozoic Era; but a few groups survived, and their descendants became common again during the Mesozoic. True coral reefs did not develop, however, until well into Tertiary times, about 60 million years ago.

Rise and Fall

We have seen how various groups of marine invertebrates have come and gone during the last 600 million years: how they have arisen, flourished for a while, and then suffered setbacks and extinction. They have not all disappeared yet, but it does seem that extinction is the eventual fate of all animal groups. The time span occupied by the different groups has varied a great deal, but the overall pattern of progress has been remarkably similar in them all. The earliest stages in the evolution of the various groups have normally been very rapid in geological terms, with the new animals spreading out into all the available habitats in a burst of adaptive radiation. Different offshoots of the original stock become adapted for life in different habitats and the rate of evolution then slows down. If conditions remain more or less constant, there is no need for the animals to change very much. The few 'living fossils' that survive today nearly all come from undisturbed and unchanging regions, such as the deep sea or remote islands. But the environment is not usually constant and in most places there have been enough slow changes to provoke the continuous evolution of animal life. There have also been some more rapid changes in the environment, and many animals have been unable to adjust to such changes. They have therefore become extinct. The animals that have suffered most in this way have been the more specialized varieties. Their more versatile, and often less elaborately constructed relatives have often been able to survive a crisis and to spread with renewed vigour afterwards. This clearly happened with the corals.

A sea urchin fossil (Stomechinus) nestling in a piece of Jurassic limestone. The limestone itself is known as an oolite, and it is composed of myriads of rounded grains. Unlike modern reefs, which are composed mostly of corals, the ancient reefs were made up of all kinds of animals.

ANCIENT REEFS

We tend to think of reefs as thick masses of limestone built up mainly from the dead skeletons of corals. The Great Barrier Reef of Australia and the coral atolls of the Pacific Ocean are typical modern examples. Corals normally comprise only about half of the building material. Huge numbers of calcareous algae also live on the reefs. Their dead skeletons fall into the crannies between corals and help to cement the fragments together. Bryozoans, crabs, sponges, molluscs, echinoderms and many other animals make their homes here, thus creating some of the most complex living communities in the world.

Reefs also existed in the remote past, but corals formed no more than a small part of these ancient reefs and were often absent altogether. The small reefs which grew in the early Cambrian seas were composed almost entirely of extinct animals called archaeocyathids. These were small conical creatures whose bodies were perforated like colanders. They were probably related to the sponges. The sponges themselves also formed large reefs in Jurassic and Cretaceous times. The archaeocyathids died out before the end of the Cambrian period, and a group of organisms called stromatoporoids took over as the dominant reef-builders. These creatures formed anvil-shaped masses of hard limestone which were nine to twelve metres (30 to 40 feet) high. We do not really know what the stromatoporoids were, but they were certainly very important reef-builders in Ordovician and Silurian times. Corals, brachiopods, molluscs, and many other animals thrived among the limestone masses.

The largest of the Palaeozoic reefs were those made by calcareous algae. Flourishing from Devonian to Permian times, these reefs were sometimes 1000 metres (over half a mile) thick. They were formed from layer upon layer of sticky, lime-secreting algae which trapped lime and mud on their surfaces. Like modern reefs, they had steep faces on the outer sides and more shallow slopes on the landward sides. Some were so hard that they have resisted erosion and still stand out as ridges on the landscape today (see page 56). The best known of these algal reefs are the Permian reefs of north-east England and of Texas.

Conflict in the Waters

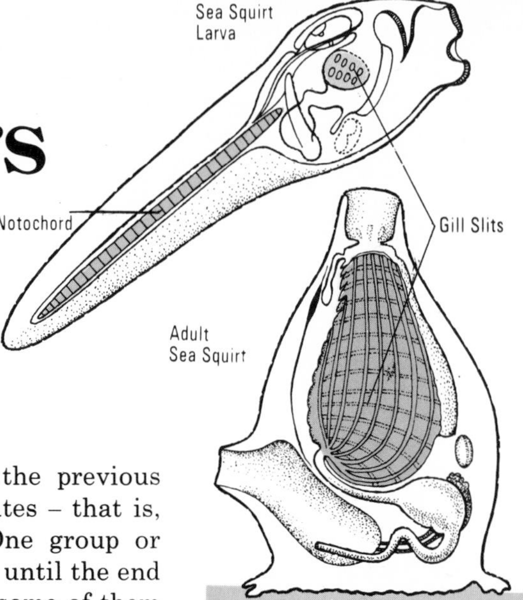

UNKNOWN ANCESTORS

The vertebrates, or backboned animals, belong to the group called the chordates. This name is taken from the *notochord*, a flexible rod that is found in no other group of animals. It runs the whole length of the body, but very few chordates retain it throughout their lives. In the vertebrates, which make up the bulk of the chordates, the notochord is usually present only in the very early stages. It is quickly replaced by the bony or cartilaginous vertebrae which make up the backbone.

It is impossible to tell from the fossil record what the first chordates looked like, or when they first appeared in the seas. These early chordates had no bones and we may never find fossils of them. It is not even certain what the earliest vertebrates were like. The earliest *known* vertebrate fossils are fish fragments from the Ordovician rocks, but these are of fishes which were already well endowed with bony armour. It is clear that the fishes had been in existence for a long time before they appeared in the fossil record.

Our best way of finding out something about the fishes' elusive ancestors is to examine the most primitive chordates alive today. These little creatures, the sea squirts and the fish-like lancelet, or *Amphioxus,* may well resemble the ancient fore-runners of the fishes. The sea squirts are bag-like animals which attach themselves to the sea bed and strain food particles from the water. They look nothing like vertebrates, but they are chordates because, among other things, their tiny larvae possess a notochord. These larvae are like little free-swimming tadpoles. It seems that swimming chordates might have arisen from such larvae by the process of neoteny or paedogenesis (see page 30). This is the process whereby young animals remain in their larval stages but can nevertheless reproduce. The lancelet may have evolved in this way from a larval sea squirt. The lancelet is a transparent, blade-like animal up to five centimetres (two inches) in length. It has a notochord and a small tail region, and, although it has neither bone nor cartilage, it is not too difficult to imagine such a creature evolving into a small fish. The vertebrates might well have evolved from a very close relative of the lancelet.

The animals described in the previous chapter were all invertebrates – that is, they had no backbones. One group or another dominated the seas until the end of the Silurian Period, but some of them then began to decline and several of them became extinct before the end of the Palaeozoic Era. The main reason for this decline was undoubtedly the arrival of the fishes – the earliest backboned animals. They took over the seas in Devonian times and, with the possible exception of part of the Jurassic Period, they have ruled there ever since. Fish also dominate the world's freshwater habitats and their descendants, in the form of mammals and birds, rule the land and the skies.

Above: Can this jelly-like creature be one of our ancestors? It does not seem very likely, if we look at the adult sea squirt. But the larva is more like a tadpole, and a form like this could have evolved into the fish-like lancelet.

Below: Lancelets, with their tails buried in the sand, filter food from the water. Muscle blocks like those of fishes can be clearly seen.

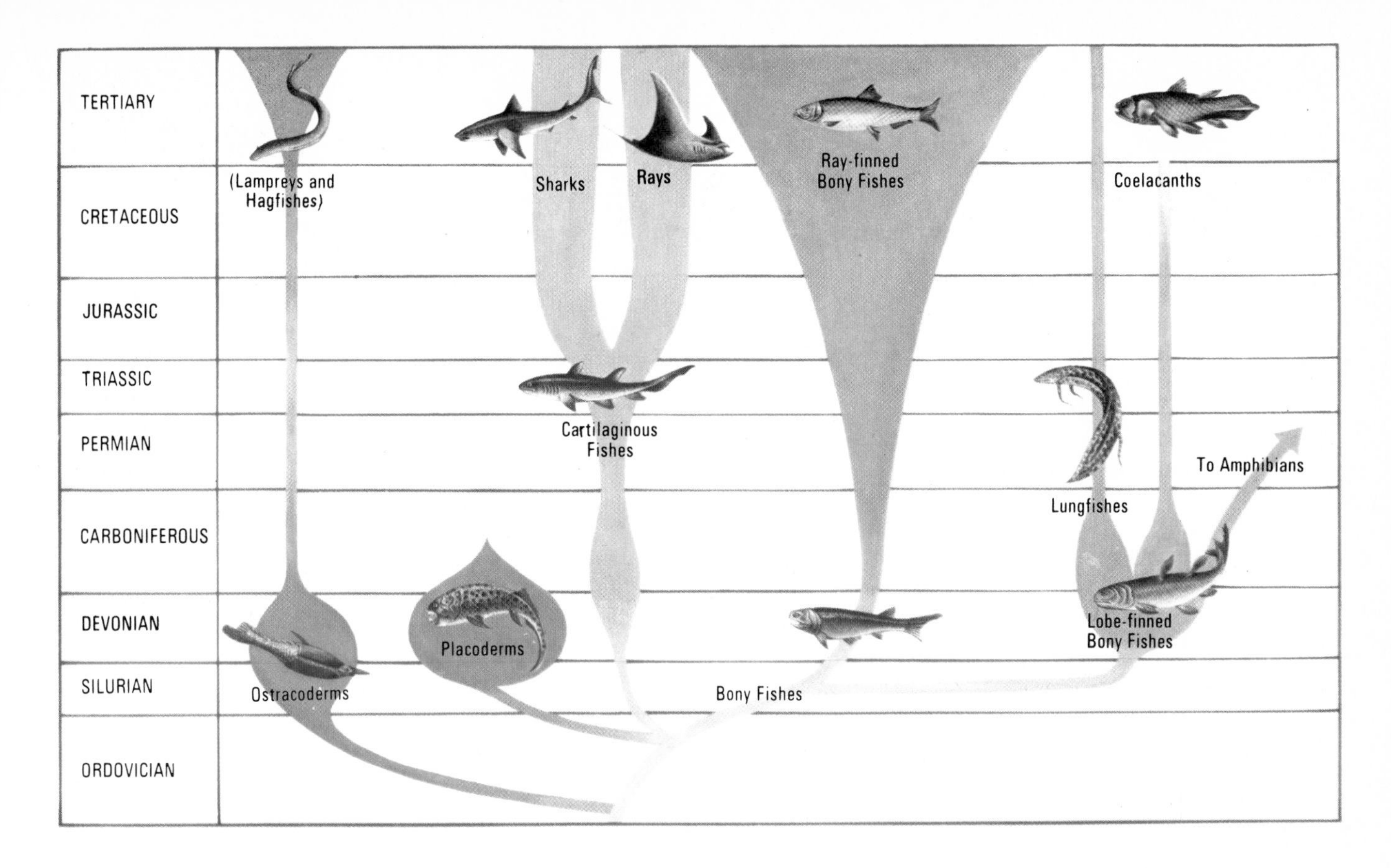

Above: Chart showing the evolutionary history of the main groups of fishes from their origins way back in the Palaeozoic Era. Until more fossils are found, we can only guess at the connexions between each group.

The Age of Fishes

The remains of a few fishes have been found in Ordovician and Silurian rocks, but it is in the Devonian layers, such as the Old Red Sandstone of Scotland and the Cleveland Shales of Ohio, that fossil fishes first appear in abundance. The Devonian Period (395–345 million years ago) is therefore known as The Age of Fishes. As well as being numerous, the Devonian fishes were very varied. Many were weirdly shaped, and most of the early ones were heavily armoured – probably as a protection against the giant scorpion-like eurypterids (see page 63) which preyed on them. All four major fish groups – the ostracoderms, placoderms, chondrichthyans, and osteichthyans – had appeared by the end of the Devonian Period. These groups are very distinct and we still do not know how they were related to each other. They had obviously passed through a long period of evolution before they appeared in the fossil record.

Jawless Boneheads

The ostracoderms were bizarre little fishes, mostly less than 50 centimetres (20 inches) long. They had no jaws – their mouths were mere holes or slits through which they sucked in water and food particles as they scavenged in the mud. Three main groups of ostracoderms are known in the fossil record. The cephalaspids were bottom-dwelling creatures with a flattened head shield composed of a single piece of bone. The eyes were on the top of the head, and the shield bore three sensory areas which probably detected vibrations in the water. The lower surface of the head was covered with small bony plates, but the floor of the mouth remained flexible. Water and mud from the sea or stream bed could thus be sucked in. The water then passed out through a number of small gill openings, but food particles were retained and swallowed. A scaly body and a tail which turned up at the end emerged from behind the armoured head. Though sluggish; the cephalaspids could obviously swim about in the water.

The second group of ostracoderms are the anaspids, most of which are found in Upper Silurian rocks. These were small fishes with tiny mouths. They had no bony head shield, but many of them were armed with rows of heavy scales. The tail turned down at the end and would have raised the head when the fish was swimming. It has therefore been suggested that these

Below: Diagrams showing how jawless fishes (top) might have evolved jaws through modification of some of the bones which supported the gills.

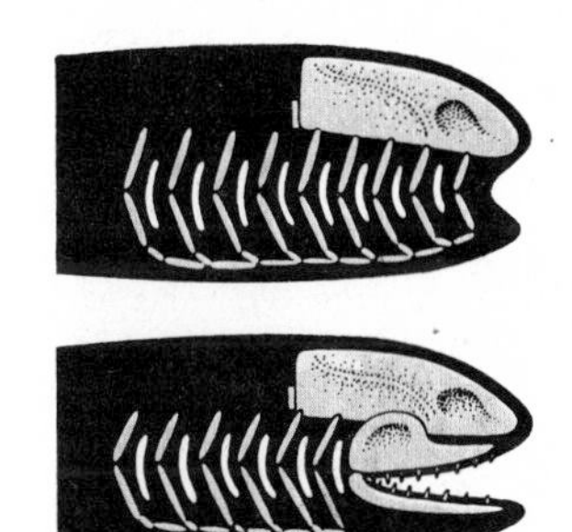

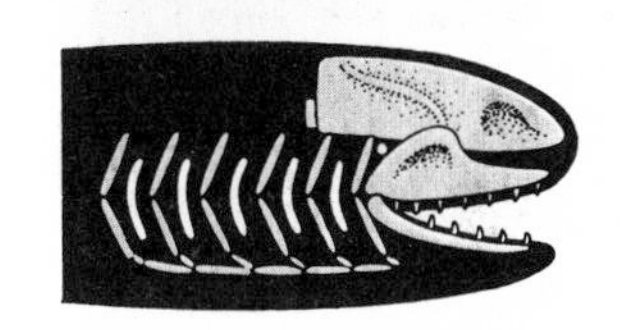

PROBLEM SOLVED

The cephalaspids puzzled zoologists throughout the 19th century. They seemed so different from anything alive today that nobody was sure whereabouts in the animal kingdom they belonged. It was even suggested that they were armoured salamanders which had returned to the water. The mystery was solved in the 1920s when Erik Stensiö, a young Swedish palaeontologist, studied a collection of Devonian cephalaspids from Greenland and Spitzbergen. By using fine needles, he managed to remove the fossilized bones of the head, leaving the rock which had formed in the cavities and passages that had once held the brain, nerves and blood vessels of the head. In this way, Stensiö was able to describe their anatomy in wonderfully complete detail and to demonstrate that the animals were primitive fishes.

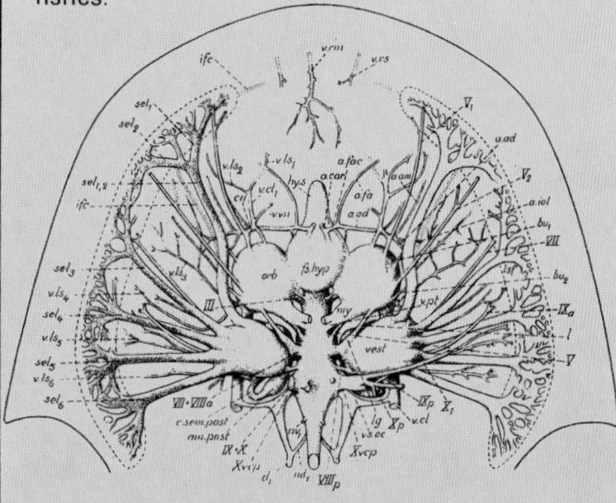

A detailed drawing by Stensiö, showing the internal structure of a cephalaspid's brain.

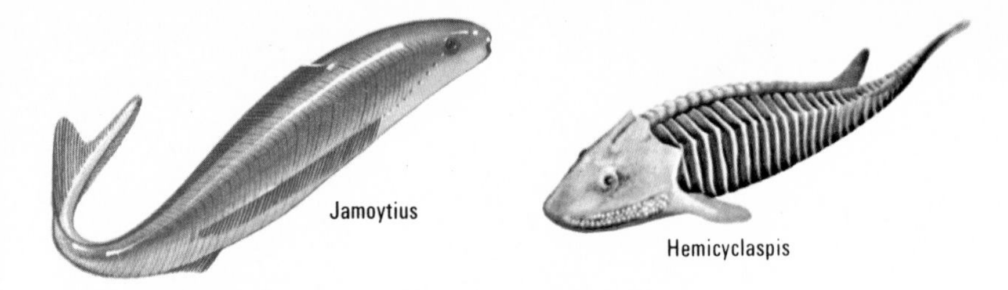

Above: Members of the three main groups of ostracoderms. Jamoytius was an anaspid with a tiny mouth. It may have used its mouth to hold on to surfaces, while its teeth scraped off the encrusting algae. The asymmetrical tail would then have driven the fish along the bottom like a plough.

Hemicyclaspis was a thick-scaled cephalaspid, which probably also fed on debris sucked up in the mud.

Drepanaspis was a bottom-living pteraspid, while the round-bodied Pteraspis probably kept clear of the bottom. Its eyes were on the side of its head, unlike those of Drepanaspis.

Thelodus does not fit easily into any of the main groups, and it may have been merely a young form of one of them.

fishes swam near the surface and fed on plankton. But this is unlikely. With their small mouths, the anaspids were probably mud-feeders like the cephalaspids. It is possible that they fed with their heads buried in the mud and their bodies almost vertical.

The anaspids, which include *Birkenia* and *Jamoytius*, are probably related to the modern lampreys and hagfishes. These are eel-shaped creatures which have lost all their bone and now have skeletons of cartilage. Although they have no jaws, their mouths hold batteries of horny teeth with which they rasp the flesh of living or dead fishes. The ancestral lampreys and

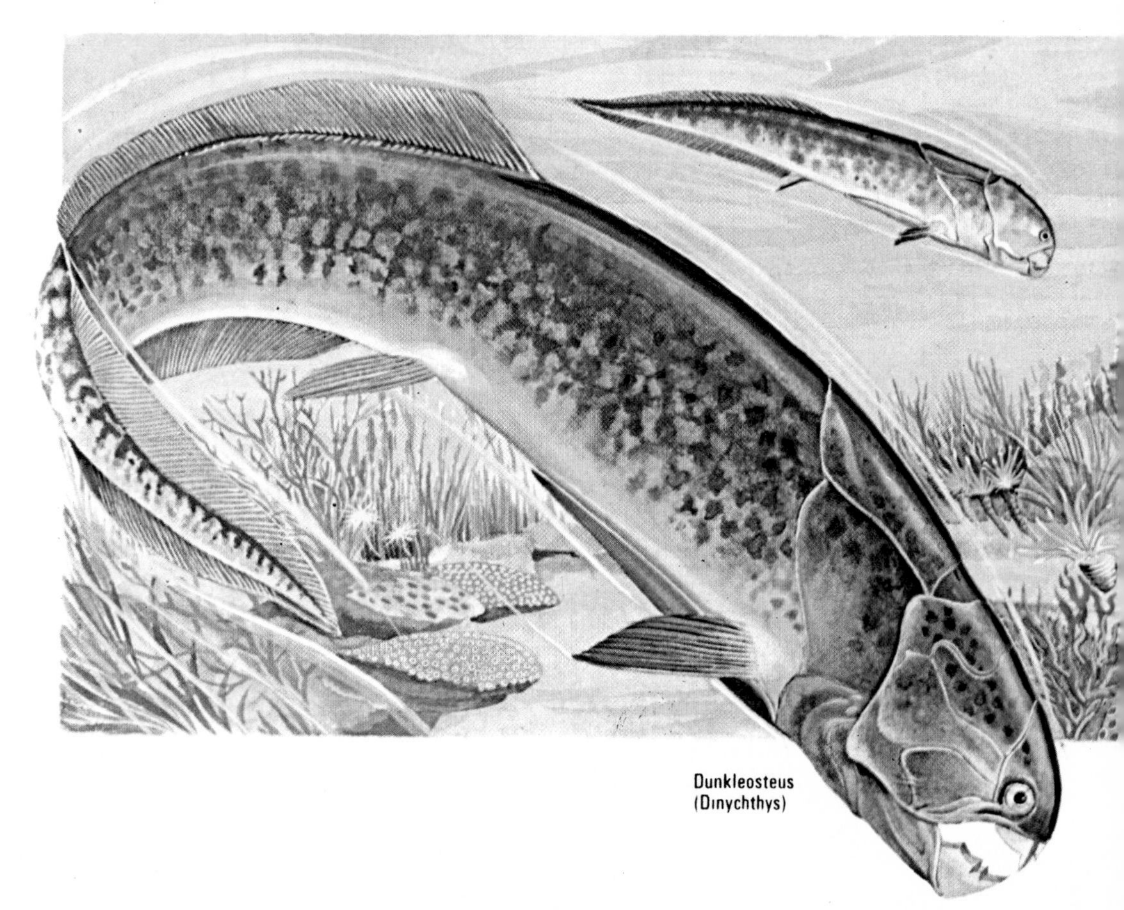

Right: The Devonian seas were ruled by the giant arthrodires such as Dinichthys (Dunkleosteus). These fearsome predators must have dined well on the small sharks known as Cladoselache, whose fossils are very numerous in the Cleveland Shales (Upper Devonian) where the Dinichthys remains have been found. Aellopos (inset right) was the ancestor of living rays. It appeared in the Jurassic and lived on the sea bed, feeding on the molluscs and other invertebrates living among the corals and seaweeds.

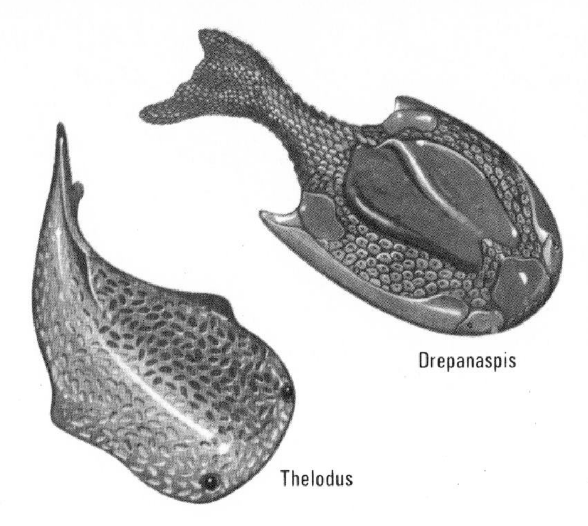

hagfishes were the only jawless fishes to survive the end of the Devonian Period: their specialized habits enabled them to compete effectively with the jawed fishes.

The third group of ostracoderms are the pteraspids. These are actually the first to appear in the fossil record, although they are not the most primitive. Like the cephalaspids, they had a head shield, but it was made up of several large plates.

Fearsome Predators

The earliest *known* jawed fishes are little creatures from the Upper Silurian rocks. We do not know when the first jawed fishes appeared, however, nor who their ancestors were. They might have evolved from some kind of pteraspid, or they might be an independent line descended from a lancelet-like ancestor. What we do know is that the Devonian waters were full of jawed fishes, many of which were active predators. Perhaps the most fearsome of all were to be found among the placoderms. These fishes had a bony covering over the head and the front part of the body, and there was a movable joint between the two sets of armour. This meant that the head could be rocked back on the neck, and the mouth could be opened very wide. The jaws had no teeth, but their jagged edges

Above: The lamprey is one of the few jawless fishes living today and it is probably a direct descendant of the Devonian anaspid fishes. Adult lampreys generally attack other fishes and feed on their blood. There is little competition for this way of life, and this may be why the lampreys have survived almost unchanged since Carboniferous times.

made them formidable weapons. The hind part of the placoderm body was unprotected and, as a rule, it was not even covered with scales. Thrashing from side to side, it could build up great power, pushing the fish easily through the water.

The placoderms were nearly all bottom-living fishes and most of them were rather flattened, but they exploited many different ways of living on or close to the bottom. The two main groups of placoderm fishes were the antiarchs and the arthrodires. The antiarchs were all fairly small, flattened fishes with weak mouths. Their eyes were situated on the top of the head, enabling them to look out for danger; presumably they led the same sort of mud-grubbing life as the jawless cephalaspids. The strangest feature of the antiarchs was their front fins. These were encased in crab-like arms, and the animal must have used them to drag itself along by 'rowing' in the mud. No other vertebrate has its limbs encased in an external skeleton. One antiarch, *Bothriolepis* from the late Devonian of Canada, has been found in such a good state of preservation that some of its soft parts are known. It had a spiral intestine like that of the shark.

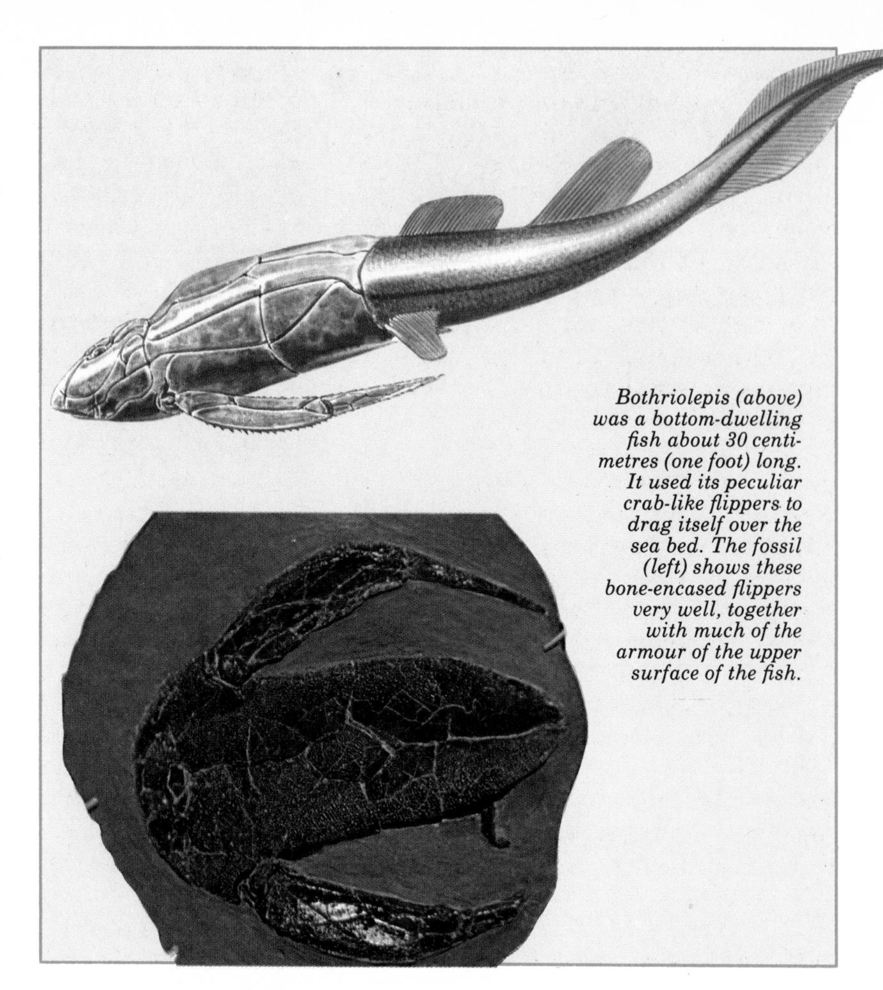

Bothriolepis (above) was a bottom-dwelling fish about 30 centimetres (one foot) long. It used its peculiar crab-like flippers to drag itself over the sea bed. The fossil (left) shows these bone-encased flippers very well, together with much of the armour of the upper surface of the fish.

Armoured Giants

The arthrodires are the largest group of placoderms. They also include the largest individual species, although not all arthrodires were large. The most primitive kinds, such as *Actinolepis*, were fairly small, flattened scavengers which lived on the bottom. These early forms were protected by a relatively long body shield. They evolved or radiated in several directions. Some remained bottom-dwellers and became highly specialized for this mode of life. *Titanichthys* was a large example of this type, reaching a length of more than two metres (six feet). Other arthrodires developed shorter armour and became predators, lurking on the bottom or else hunting in mid-water. Some of these predators also grew very large. *Dinichthys* (*Dunkleosteus*), from the Cleveland Shales, was over nine metres (30 feet) long.

The placoderms were very numerous and apparently very efficient during Devonian times, living in the seas and in fresh water. But they disappeared completely during the Carboniferous Period. Their place was taken by the sharks and rays.

Below: A reconstruction of the massive, armoured head and jaws of Dinichthys (Dunkleosteus). The human head, drawn to the same scale, shows just how huge this fish must have been. The enormous jagged jaws formed a lethal trap for other fishes. They were powerful enough to crunch right through the armour of the antiarchs and ostracoderms.

Skeletons of Cartilage

The chondrichthyan fishes, which include the sharks and rays and their relatives, have lost all traces of bone: their skeletons are made entirely of cartilage. This is a tough, gristly material which is often strengthened with lime. Since they have no bone, these fishes have not been fossilized in large numbers. The most common remains are their triangular or dagger-like teeth (see page 17). Many early sharks are known only from their teeth, but one primitive type has been very well preserved in the Cleveland Shales of Ohio. Known as *Cladoselache*, it lacked the armour of the placoderms and it was very much like a modern shark in appearance. It was about one metre (three feet) long and its paired fins were simple triangular flaps. Rows of sharp teeth show that it was a predator. Fossils have even been found showing the teeth and scales of other fishes in the stomach. *Cladoselache* was clearly a fast-moving active hunter like today's sharks, but with one important difference. Modern sharks have small eyes and they hunt almost entirely by smell, whereas *Cladoselache* had large eyes and a short snout. It must have hunted mainly by sight.

Although the Palaeozoic sharks were similar to those of today, the two groups are not closely related. During the Carboniferous Period some sharks evolved flattened back teeth suitable for crushing molluscs and other hard-shelled animals. They began to visit the sea bed, which had previously been occupied by the placoderms. Possibly because they had two kinds of teeth, and could therefore eat two kinds of food, these sharks managed to survive the upheavals which caused so many animals to become extinct at the end of the Permian Period. Sharks with the two types of teeth were common in Triassic and Jurassic times. *Hybodus* is a well-known example. Towards the end of the Jurassic Period, however, there was another burst of evolution. The hybodontoid fishes diverged into two main groups. One of these gave rise to the modern sharks and the other to the skates and rays. *Hybodus* and most of the other 'mixed toothed' sharks became extinct, but a few, such as the Port Jackson shark of tropical seas, still survive.

Modern sharks are nearly all predatory fishes with rows of razor-sharp teeth. Some are very dangerous to man. Others, however, have become adapted for feeding on plankton. The 18-metre (60-foot) whale shark, the largest living fish, is one of these, and so is the 14-metre (45-foot) basking shark.

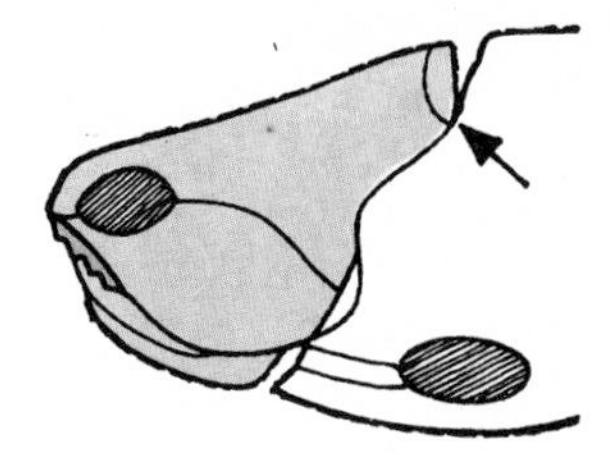

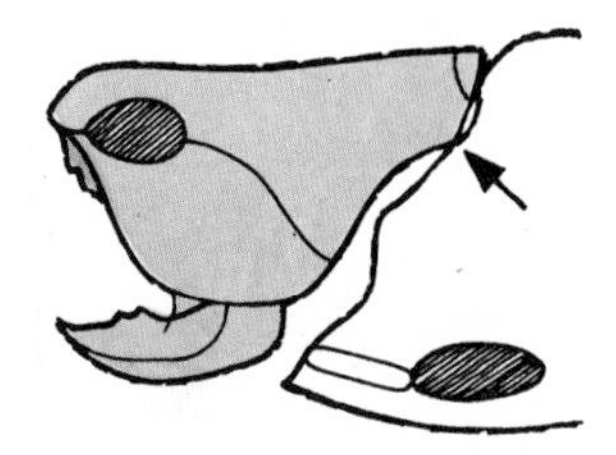

Above: The head and trunk armour of a Devonian arthrodire, showing how the head (blue) was pivoted on the trunk armour. The head could thus be raised and the gape of the mouth made very wide. Notice the large eyes of this predatory fish.

Below: Xiphotrygon, an extinct sting ray from the Eocene rocks of North America.

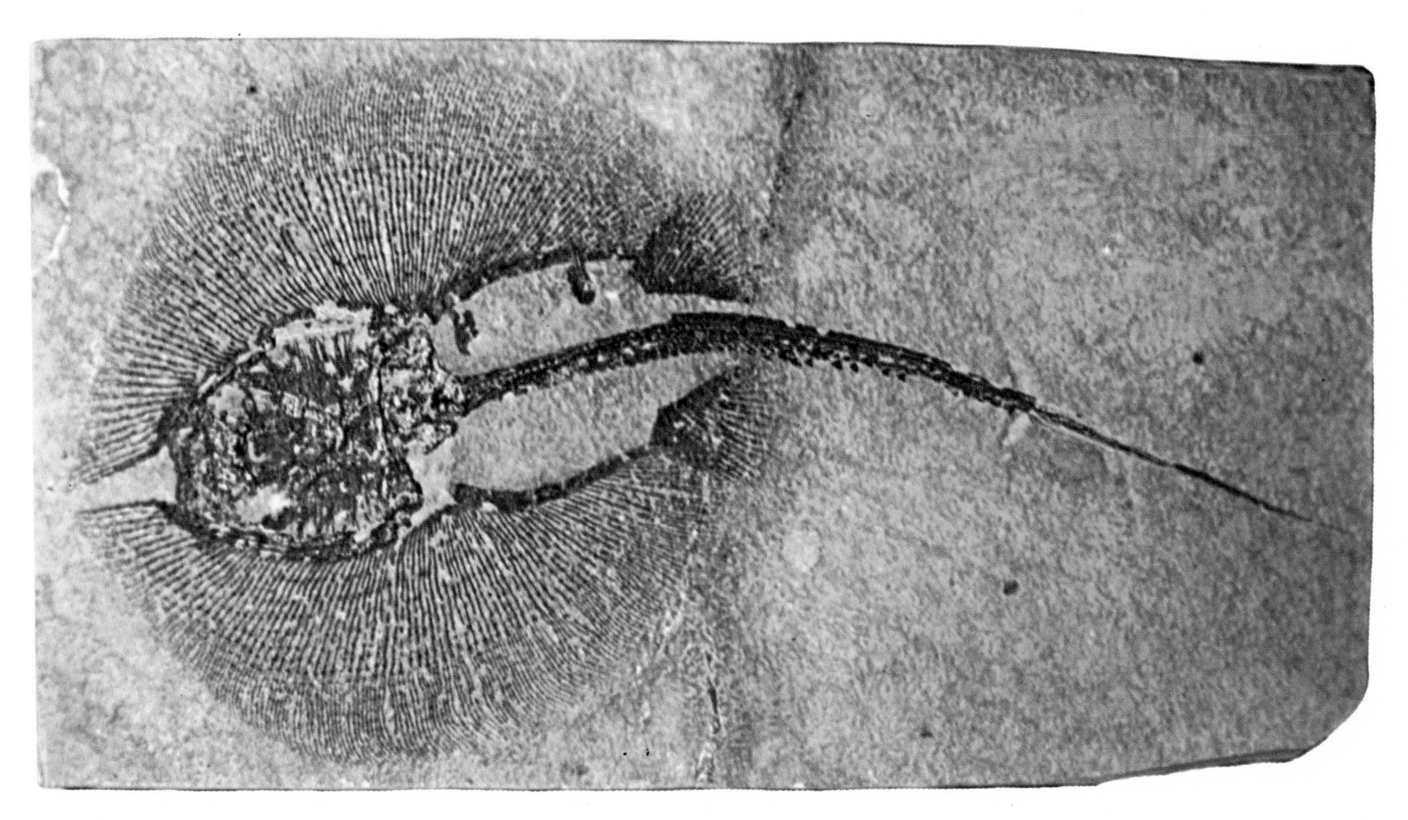

Above: A model of *Eusthenopteron* showing how this ancient rhipidistian probably used its fins to support some of its weight. Notice the large scales and the bony armour on its head.

The Lobe-Finned Fishes

Although there are many cartilaginous fishes living today, they are greatly outnumbered by the osteichthyans, or bony fishes. These are of two main kinds: the lobe-fins, most of which are extinct, and the ray-fins. The lobe-finned fishes are of particular interest because the first land-living vertebrates, the amphibians, evolved from them. The lobe-fins take their name from the fact that their fins are carried on scale-covered, muscular lobes. These lobes are like short, stubby limbs, and it is from these that the legs of the land-living vertebrates evolved.

Below: Two ancient lobe-finned fishes. The fleshy lobes are clearly seen, especially in the paired fins which eventually evolved into limbs. *Gyroptychius* (bottom) was a rhipidistian, *Griphognathus* was an early lungfish.

The Rhipidistians

There are three main groups of lobe-fins. The group which gave rise to the amphibians was that of the rhipidistians, of which *Eusthenopteron* is the best known. The rhipidistians were common in Devonian times, and a few survived until the end of the Palaeozoic Era. They reached three metres (ten feet) or more in length and they were flesh-eaters. Lurking in shallow waters, both fresh and salt, they probably used their stout fins to hold their bodies clear of the bottom. Like many other Devonian fishes, the rhipidistians had lungs and could breathe air if the water became foul and poor in oxygen.

The earliest rhipidistians, such as *Osteolepis* from the middle of the Devonian, were slender fishes covered with heavy scales. The head region was protected by bony plates. Later rhipidistians lost much of their bony armour and their scales became much thinner.

Coelacanths

Closely related to the rhipidistians are the coelacanths, which first appear in the Middle Devonian rocks. The coelacanths are rather deep-bodied fishes with a more or less symmetrical, three-lobed tail. They

The Australian lungfish (Neoceratodus) in the photograph has changed little from the Devonian lobe-fins and it probably lives in the same way that they did. It has lost the heavy armour from the head, however. The Australian lung-fish is the most primitive of the living species and cannot survive out of water for very long. The African (Protopterus) and South American (Lepidosiren) lungfishes can survive drought by curling up in dug-out chambers in the mud of a drying river bed. Fossil burrows of this type are known from Permian rocks, which proves that ancient lungfishes had the same habit.

differ from the rhipidistians in that their nostrils do not open into the back of the mouth. In the coelacanths the lung has become a true swim bladder and cannot be used for breathing. In some deep-water forms, such as the Mesozoic *Macropoma,* the swim bladder had bony walls, presumably to prevent it from collapsing under pressure.

Lungfishes

The third group of lobe-fins are the lungfishes or Dipnoi. The primitive Devonian lungfishes, such as *Dipterus*, were covered with thick, round scales. Their fins were similar to those of the other lobe-finned fishes. The head was protected by numerous bony plates, and the upper jaws were firmly fixed to the rest of the skull. The teeth, in the form of crushing plates, could inflict a crippling bite. Many varieties of lungfish lived in the fresh waters of the Devonian and Carboniferous Periods. Like the rhipidistians, they could breathe air into their lungs if the water became foul.

The lungfishes become rare in the fossil record towards the end of the Palaeozoic Era, but three genera are still living today.

COELACANTHS ALIVE

On 22nd December, 1938, a fishing boat caught a strange fish off the east coast of southern Africa. The curator of the local museum came to see it, and later sent a sketch of it to Dr J. L. B. Smith, an expert on fishes. On seeing the sketch, Dr Smith said that he would not have been more surprised if he had seen a dinosaur walking down the street. The two back fins and the fleshy paired fins meant that it was a lobe-fin, and the shape of the tail showed that it was a coelacanth – a member of a group which, at that time, was thought to have been extinct for 65 million years. Smith named this living fossil *Latimeria* after Miss Courtenay-Latimer, the curator who sent him the sketch. About 50 more coelacanths have since been caught.

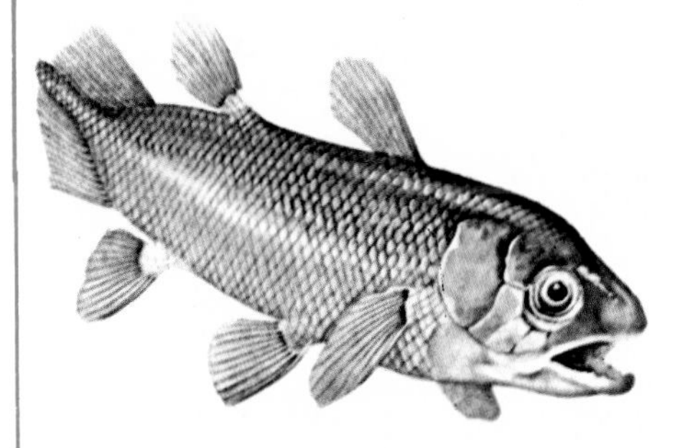

The coelacanth, Latimeria, is about one and a half metres (five feet) long. It weighs about 70 kilogrammes (150 pounds).

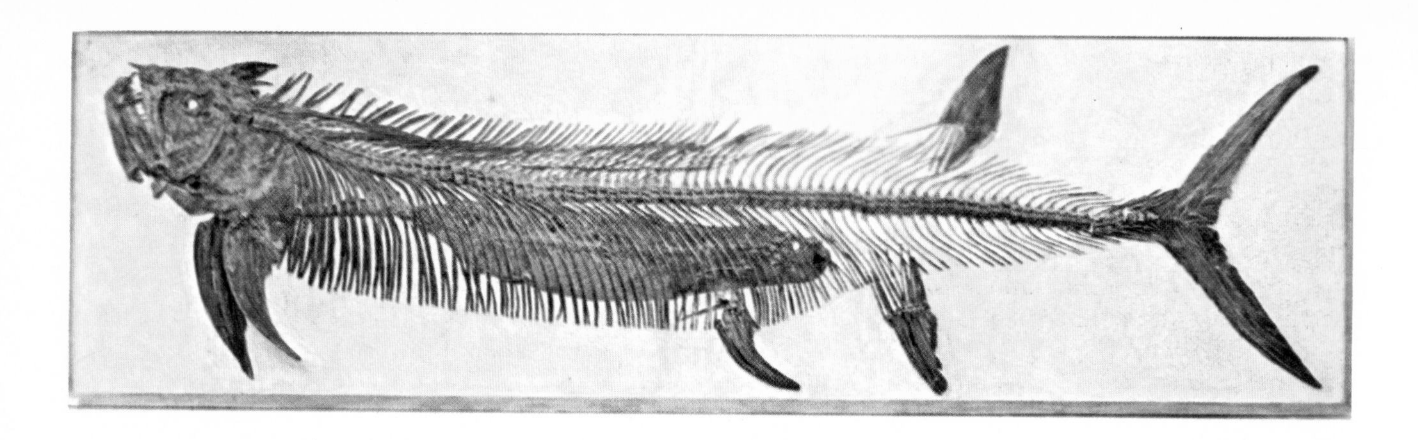

Above: Two fossils for the price of one: this Xiphactinus (Portheus), a giant Cretaceous fish which reached lengths of about five metres (16 feet), must have eaten another fish just before it died, because the smaller fish has been perfectly preserved inside it. Found in Kansas, this specimen is the largest known fossil teleost.

Rulers of the Waves

The fresh and salt waters of today are dominated by the ray-finned fishes. This is the largest group of vertebrates alive today – it includes over 20,000 species. The ray-fins are a very ancient group, however, for they made their first appearance as far back as the Devonian Period. Their fins have no lobes but are supported by a number of bony rays which grow straight out from the body. The ray-fins, also known as actinopterygians, fall into three main groups which more or less follow one another chronologically.

The earliest group of ray-fins were the chondrosteans, and the earliest known member of the group is *Cheirolepis*, from the Old Red Sandstone of Scotland. *Cheirolepis* was about 23 centimetres (nine inches) long, and was covered with thick scales. It was an active predator in fresh water. The chondrosteans remained as rather insignificant freshwater fishes for much of the Palaeozoic Era, but towards the end of that time they began to change. The placoderms and other ancient groups were fast disappearing, and the ray-fins seized their opportunity. They invaded new habitats, including the sea, and began to develop useful adaptations.

One major change was in the head and jaws. The early chondrosteans, like their fellow placoderms and lobe-fins, had had bony armour on their heads. Although this gave them good protection, it restricted the movement of the jaw muscles, and the jaws were not particularly efficient. The gradual reduction of the head armour allowed the fishes to develop mouths which could be extended to snap at prey and suck in small animals. There were also changes in the body and tail. The scales became much thinner, and the primitive lung evolved into a swim bladder. By altering the amount of gas in the bladder, the fishes could control their buoyancy and float effortlessly at any depth. This meant that the tail no longer needed to provide lift, and as a result it gradually became symmetrical. The body became shorter and deeper, and the paired fins also became more manoeuvrable.

Cornuboniscus, a sprat-sized chondrostean from the Carboniferous Period.

Cheirolepis, from the Middle and Upper Devonian is the earliest known chondrostean.

Caturus, from the Jurassic, was a much more streamlined fish, which could have given rise to the modern fishes. Notice that Caturus had a symmetrical tail.

Holosteans and Teleosts

The first fishes to undergo these changes were the holosteans, the second major group of ray-fins. They first appeared in late Permian times, and during the Triassic Period they replaced the chondrosteans

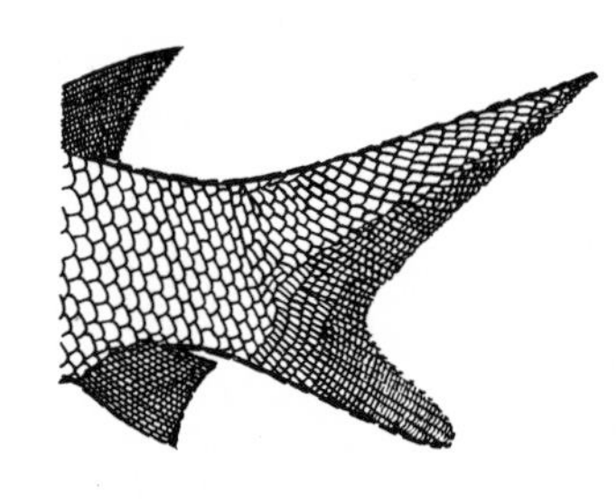

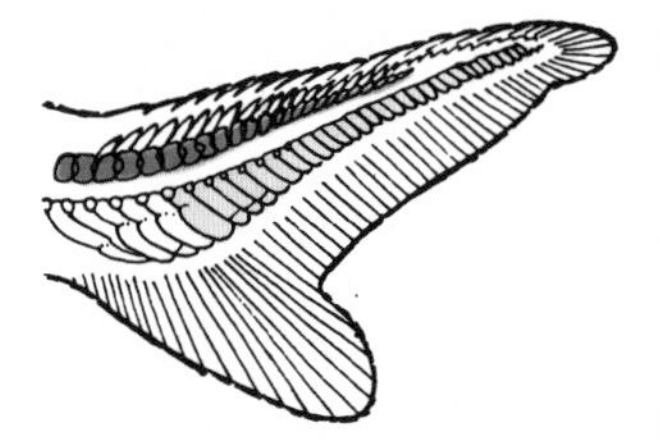

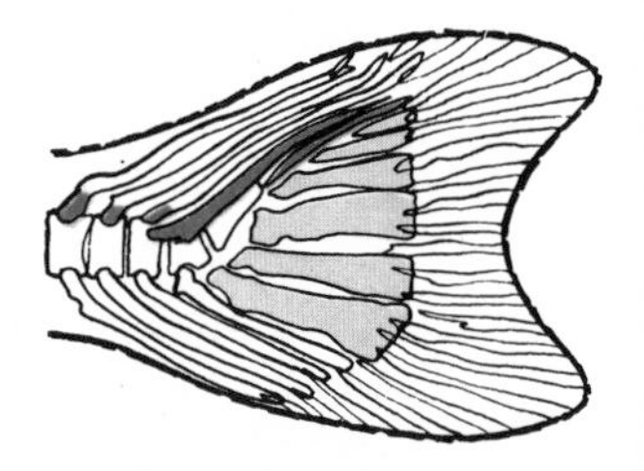

Above: The up-turned, or heterocercal, tails of an ancient chondrostean (top) and a sturgeon (middle), compared with the tail of a teleost. The scales have been removed from the two lower tails to show how symmetry has been attained by differential growth of the tail bones.

almost completely. Two or three chondrostean species still survive, however. These are the sturgeon, the peculiar paddle-fish *(Polyodon)* of the Mississippi, and possibly the bichir *(Polypterus)* of Africa. No-one is quite sure where the bichir fits into the evolutionary story, but it certainly has the large scales of the chondrosteans. It is interesting to notice that, apart from the sturgeon, these ancient fishes have survived only in restricted areas of fresh water where there is relatively little competition.

The holosteans may have evolved from several groups of chrondrosteans, but when once they had appeared they underwent a marked radiation (see page 31) and evolved into a wide variety of types during the Triassic and Jurassic Periods. Later, one of these types – possibly a streamlined predator like *Caturus* – gradually gave rise to the third major group of ray-fins, the teleosts. Only two holosteans have survived to the present day. These are the bowfin *(Amia)* and the garpike *(Lepisosteus)*, both of which have survived only in certain North American rivers. The garpike still has scales so thick that an under-water spear will bounce off its body.

The teleosts first appeared in the Jurassic Period. The new features which had characterized the holosteans had now become even more pronounced, giving the teleosts very thin, rounded scales, fewer bones in the skull, and a perfectly symmetrical tail. It was not until late in the Cretaceous Period that the teleosts really began to replace the holosteans, but after that the change-over proceeded very quickly. The teleosts evolved into many lines and adopted a wide range of habits. They became the rulers of the world's waters and have remained so up to the present day.

This beautifully preserved fossil teleost comes from the Green River in North America. It lived during the Eocene, and in life must have resembled the elegant spiny-finned fishes that live today in tropical seas.

The Conquest of the Land

Life began in the water and was confined to it for many millions of years. Plants and animals have now conquered the land as well, but they are still very much dependent on water for their existence. Although they may contain hard parts, such as bones or wood, their bodies are composed very largely of water. If they lose much of this water, they die. Breathing or respiration also depends on water, because oxygen can pass into the body only through a moist surface.

In adapting to life on land, plants and animals had therefore to evolve waterproof skins which retained moisture in their bodies. At the same time they had to evolve ways of keeping their breathing surfaces moist without letting too much water evaporate. They also had to be able to get rid of their waste products with a minimum loss of water. Aquatic creatures have no problems with their poisonous wastes, because there is plenty of water to dilute them and wash them away. Land-dwellers usually produce less-poisonous waste products, so that they need less water to dilute them to a safe level. Some convert their wastes into solid materials and lose very little water in getting rid of them.

The land-living organisms also had to be able to support their own weight. Their aquatic ancestors did not have to face this problem because most of their weight was supported by the surrounding water. Finally, the land-dwellers had to cope with much greater variations in temperature than their aquatic forebears. This is because air warms up and cools down much more rapidly than water.

With all these difficulties confronting them, it may seem surprising that any organisms should ever have ventured out of the buoyant waters. In fact, if these physical factors had been the only ones involved, they might well never have done so. But an organism's environment also includes its relationship with other organisms, all competing with one another for food and for homes and all striving to eat rather than be eaten. After countless millions of years of evolution in the water, most of the *niches* or modes of life open to aquatic creatures must already have been occupied. Competition was fierce, and this was just the incentive that was needed to send the plants and animals out to conquer the land.

Opposite: The first fishes to leave the water must have lived in much the same way as today's mud-skippers. These little fishes crawl on the mud at low tide, and absorb oxygen through the skin and through membranes in their mouths and gill cavities.

RESPIRATION ON LAND

Although an animal living on land must enclose its lungs or gills to prevent them from drying up, breathing is otherwise much easier on land than in the water. Keeping a flow of air moving across the respiratory surfaces demands much less energy than pumping water across them. Air also contains over 20 per cent oxygen, while the amount of oxygen dissolved in water is less than one per cent. It is thus much easier to absorb oxygen from air than from water.

Getting Ready

Certain changes had to take place, of course, before any animals or plants could move ashore. Without such *pre-adaptive changes,* the organisms would not have been able to spend even short periods on the land. But, when once they found that they could survive there for a while, natural selection came into action. Those organisms with a combination of characteristics allowing them to survive on land for the longest periods were the most successful. They were therefore favoured by natural selection, and the invasion of the land was under way.

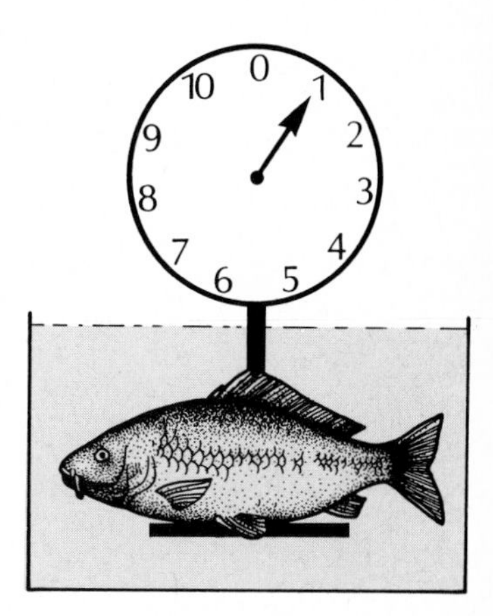

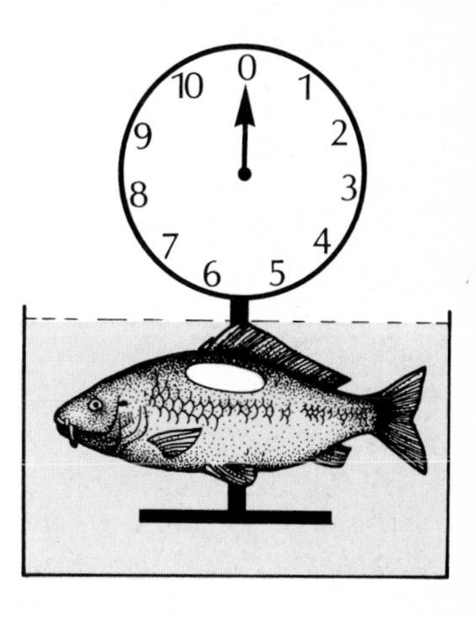

Organic matter is only slightly more dense than water. Thus a fish that weighs 10 grammes in air weighs only 0·7 grammes when submerged in water. The buoyancy, or upthrust, of the water balances out the remainder of the weight. The same fish equipped with an air-filled space equal to seven per cent of its volume can hang weightless in the water. This is the function of the swim bladder of the teleosts. For a fish venturing out of the water, the seeming increase in weight would be a considerable problem.

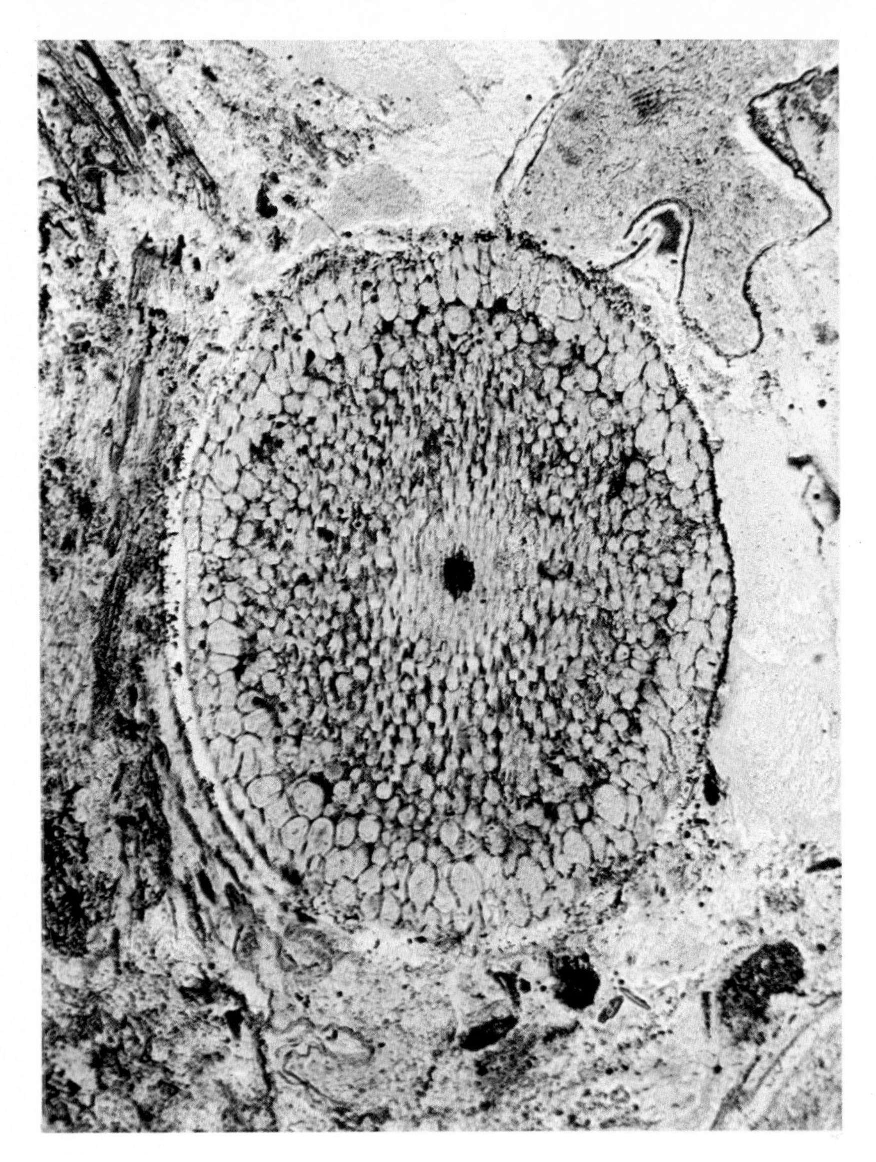

Above: Rhynia (right) was one of the earliest land plants. It had no real roots; instead, clusters of tiny hairs on the underground parts of the stems served to absorb water. There were no leaves or flowers, and the plants carried spores in the oval capsules at the top of the stems. Specimens of Rhynia are preserved in the Rhynie Chert of Scotland (left). This thin section of a stem shows very clearly the water-carrying cells in the centre and the waterproof cuticle on the outside, indicating that the plants really did live on the land.

The Plants Invade

It is clear that plants must have colonized the land before any animals did so, because there would have been no food for the animals until the plants had established themselves. As far as we can tell from the fossil record, it was in late Silurian times, just over 400 million years ago, that plants first acquired the anatomical features which enabled them to adapt to life on land. The most important of these new features was the vascular system – a system of tubular cells carrying water around the plants. Algae are usually surrounded by water and have no water-carrying system of this kind, but a vascular system is found in nearly all land-living plants, including ferns, conifers and flowering plants. The main source of water for land-living plants is the soil, and the vascular system carries the water from the roots to the aerial parts. The vascular system is clearly an adaptation for land life. It is thought to have appeared during, or soon after, the initial invasion of the land. The other major feature involved in this process of adaptation was the development of the cuticle, a waxy layer forming a waterproof coat for the aerial parts of the plants.

The earliest known example of a vascular plant is *Cooksonia*. Fossils of it have been found in Upper Silurian rocks in various parts of the world. The plant was only a few centimetres high, and was exceedingly simple – just a collection of branching stems ending in rounded spore capsules (see page 44). Running through the centre of each stem was a strand of water-carrying tissue, easily recognized by the distinctive thickenings of the cell walls. *Cooksonia* is known only from flattened carbon films in the rocks, but some slightly younger plants of a similar type are exquisitely preserved as solid fossils in the Rhynie Chert of Aberdeenshire. These plants are called *Rhynia*. They show very clearly the vascular tissues, the cuticle, and various other adaptations for life on land.

Experimental Plants

It is important to remember that not all of today's land-living plants have vascular tissues. Fungi, mosses, liverworts, and certain algae are common (although rather inconspicuous) members of modern land floras. These little non-vascular plants may well have existed on land during late Silurian times, although no fossils of them have been discovered. What we do often find in association with the earliest vascular plants are some peculiar forms which are quite unlike any living plants. A good example is *Prototaxites,* usually regarded as an alga because it cannot be fitted into any other plant group. When it was first found in the Lower Devonian of Canada, it was thought to be the fossilized trunk of a yew tree, but similar types from the Silurian rocks are much smaller. *Prototaxites* disappeared at the end of the Devonian Period – an unsuccessful experiment in the evolution of land plants.

The Elusive Pioneers

The structure of *Cooksonia* indicates that it grew on dry land. All earlier plant fossils have been identified as aquatic algae. The immediate ancestors of the vascular plants are therefore unknown. These were the pioneers, which presumably gained a foothold on the land because they had some means of reducing water loss from their tissues. The vascular plants are most probably descended from the green algae, for the two groups have the same kinds of chlorophyll and they both store food in the form of starch. In addition, they both have the same kind of cell wall construction. Some living green algae are even covered with a cuticle-like structure and show some ability to survive on land. But fossil evidence is lacking and palaeontologists continue to search Silurian rocks for the missing links.

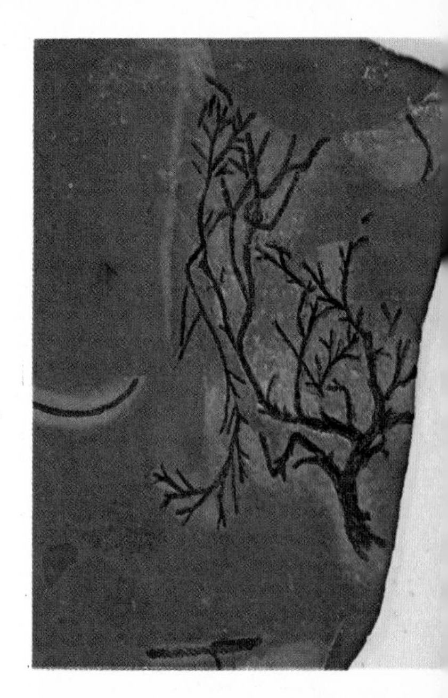

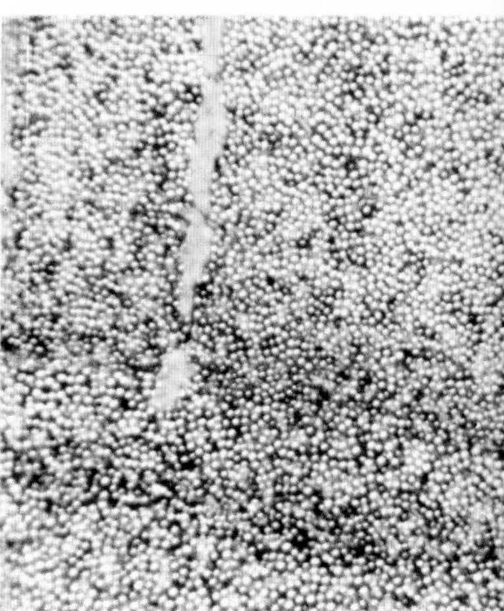

Right top: A newly discovered branching alga from Wales. The graptolite embedded in the same rock dates it as Silurian.

Right: The axis of Prototaxites cut across to show the cylindrical tubes. Palaeontologists are not sure whether it was an early vascular land plant or a marine alga.

Below: In Devonian times the hillsides were bare, for land plants were still confined to the marshy edges of lakes, as shown here. The small plants in the foreground are Cooksonia, while the larger ones, up to 30 centimetres (one foot) in life, are Zosterophyllum. The earth has been removed from an area on the right to show the tangled rhizomes from which the aerial shoots arose.

The Invertebrate Invasion

When the plants had become established on land, the way was open for the animals to follow. Like the first land plants, the first land animals must have lived in the moist habitats around ponds and streams, where they could feed on decaying plant material. These first land animals were invertebrates belonging to the large group known as arthropods. With their tough external skeletons and limbs, these animals had few further adjustments to make for life on land. A waterproof layer on the outside of the skeleton cut down the rate at which water was lost from the body. In the insects, the problem of respiration was solved by a system of breathing tubes called tracheae. These opened to the air through tiny holes in the skeleton and carried oxygen directly to every organ in the body.

One disadvantage from which arthropods suffer to this day is that an external skeleton cannot grow. Arthropods must therefore periodically moult their skeletons. This involves casting off the old outer coat and secreting a new, larger one. This system limits the size of the arthropod body.

Monster Millipedes

We do not know exactly when the arthropods left the water, but there were certainly plenty of them on the land during the Carboniferous Period. Millipedes were particularly common. The spectacular arthropleurids seem to have been related to the millipedes, although until recently they were regarded as terrestrial trilobites. Among the largest of all arthropods, they were up to two metres (six feet) long and rather flat. One specimen has been found with its gut full of clubmoss fragments, and it seems that the arthropleurids lived by ploughing through leaf litter and swallowing the debris, just as some modern millipedes do.

Above: There is little doubt that this leaf from the Gondwanaland seed fern Glossopteris had been nibbled by an insect before it became fossilized.

Below: A reconstruction of Arthropleura, a monstrous arthropod, that ploughed its way through the deep leaf litter of the coal forests.

Spiders and Scorpions

Scorpions seem to have been truly aquatic in pre-Carboniferous times, but amphibious and fully terrestrial species appeared in the Carboniferous Period. An early amphibious form from Scotland was 45 centimetres (18 inches) long. Another group of Carboniferous arachnids were the Ricinulei – a group of plump and rather primitive spiders. A few species still survive as 'living fossils' in the tropical forests, where they hide among wet leaves and under rotten logs much as their Carboniferous ancestors must have done. Like nearly all the early arachnids, the Ricinulei were carnivorous – they fed on the early insects.

Mites were also common in Carboniferous times. Their droppings have been found in association with plant fragments in coal balls (see page 48). The mites probably chewed up plant debris discarded by the millipedes. By breaking it down into very fine particles, they played an important part in the formation of soil.

The First Insects

The first true insects appeared in Carboniferous times, and winged insects were abundant in the coal forests. Cockroaches, primitive grasshoppers, mayflies, and giant dragonfly-like insects were among the common forms. Some of these early insects had already developed long piercing beaks which were presumably used to suck liquid food from the abundant and succulent clubmosses. They may also have attacked the seeds of seed ferns, for several fossil seeds have been found with neat holes drilled in them. The insects, then, were clearly dependent on the plants, and probably helped to pollinate the seed ferns even at this early stage.

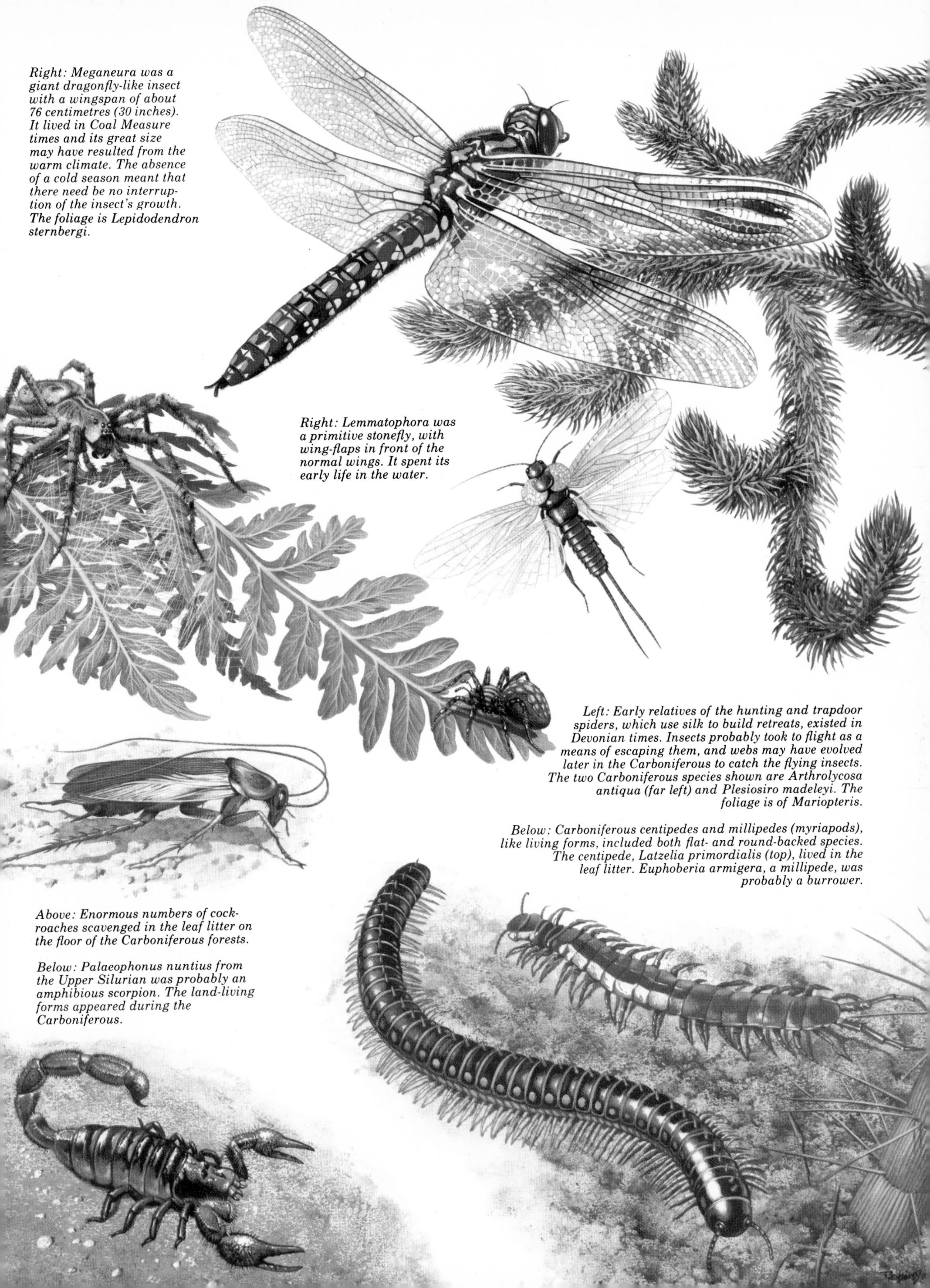

Right: Meganeura was a giant dragonfly-like insect with a wingspan of about 76 centimetres (30 inches). It lived in Coal Measure times and its great size may have resulted from the warm climate. The absence of a cold season meant that there need be no interruption of the insect's growth. The foliage is Lepidodendron sternbergi.

Right: Lemmatophora was a primitive stonefly, with wing-flaps in front of the normal wings. It spent its early life in the water.

Left: Early relatives of the hunting and trapdoor spiders, which use silk to build retreats, existed in Devonian times. Insects probably took to flight as a means of escaping them, and webs may have evolved later in the Carboniferous to catch the flying insects. The two Carboniferous species shown are Arthrolycosa antiqua (far left) and Plesiosiro madeleyi. The foliage is of Mariopteris.

Below: Carboniferous centipedes and millipedes (myriapods), like living forms, included both flat- and round-backed species. The centipede, Latzelia primordialis (top), lived in the leaf litter. Euphoberia armigera, a millipede, was probably a burrower.

Above: Enormous numbers of cockroaches scavenged in the leaf litter on the floor of the Carboniferous forests.

Below: Palaeophonus nuntius from the Upper Silurian was probably an amphibious scorpion. The land-living forms appeared during the Carboniferous.

A scene from the late Devonian. On the left is an adult Eusthenopteron. In the background, young Eusthenopteron fishes are sheltering in the shallows. On the banks, they would have found an abundant source of food – the larvae of insects and other invertebrates – and this might have given them an incentive to travel further from the water.

The adult rhipidistians, dependent on fish for their food and too heavy to move out of water, would almost certainly have been unable to make the transition to land. But the lighter young would have found it relatively easy. Thus, it seems likely that these young fishes eventually evolved into the first amphibians, such as Ichthyostega (right).

Like modern amphibians, Ichthyostega probably laid its eggs in the water and passed through an aquatic larval stage. Even the adult still had a tail fin and probably spent most of its time in the water. But it had lost its gills and must have relied on its lungs for breathing.

LUNGFISHES

The Australian lungfish (below) cannot stay out of water for long because it relies mainly on its gills for its oxygen supply. Its African and South American relatives, however, can survive even if their lakes or streams dry up completely. They burrow into the mud and each hollows out a small chamber in which it curls up. A narrow passage maintains contact with the surface, and the fish uses its lungs to obtain oxygen from the air. The fishes can survive in this state for up to two years.

Like modern amphibians, the lungfishes fill their lungs by first filling their mouths and gill chambers with air. They then use the skeleton and musculature of the floor of the mouth to pump the air back into the lungs.

From Fins to Feet

If we compare a living amphibian with one of today's typical bony fishes, the transition from water to land seems to present immense difficulties. The frail fins of the modern fish, together with its lack of lungs, would make it quite helpless on land. But, if we go back to Palaeozoic times and look at some of the extinct fishes, the difficulties nearly all disappear. As we have already seen (see page 76), the rhipidistian fishes of the Palaeozoic included a creature called *Eusthenopteron*. Like all rhipidistians, it had paired fins that contained bones apparently identical in position and function to the limb bones of the early amphibians. It also almost certainly had lungs, like those of its relatives, the lungfishes.

Eusthenopteron was therefore probably capable of surviving on land and even of moving clumsily over it. But it was a heavy fish, up to 60 centimetres (24 inches) long, and its powerful jaws suggest that it fed on other fish. Why, then, should such a fish ever venture out of the water into an environment devoid of such food?

At one time it was generally believed that such fishes left the water only because they had to. Many of the Devonian rocks in which they have been found are

red sandstones, and it was suggested that such rocks could have been deposited only in conditions of seasonal drought. Fishes such as *Eusthenopteron* might then have found their stream or pond drying up around them, and might have struggled across land in search of a more permanent body of water. But they would still have been merely tolerating land as an unpleasant interlude between meals in different ponds.

We now know, however, that red sandstones can be deposited in tropical and subtropical humid conditions even if there are no seasonal droughts. As a result, it is easy to imagine the colonization of the land taking place in a less hostile environment. It is also more likely to have been accomplished by young fishes rather than by adults.

Young fishes are preyed upon by larger fishes of all kinds, including their own, and often escape by moving into the shallows where the larger fishes cannot easily follow. If the young rhipidistians did crowd into the shallows in this way, it is not difficult to imagine them using their strong lobe-fins to drag their lightweight bodies out on to the surrounding mud. Here they would have found shelter in the vegetation, as well as a plentiful supply of insect larvae and other small invertebrates to eat. Well-fed and safe from their enemies, the young fishes would have been in no hurry to return to the water: in fact, there was every encouragement for them to stay on land. And it seems that they did so, leaving their descendants to evolve gradually into amphibians.

Below: The drawings show how the fin of a rhipidistian fish (left) may have evolved into the clumsy walking leg of a primitive amphibian (right). The major bones of the fin and the leg are identical in position and musculature. Although the intervening stages are hypothetical, changes similar to those shown must have taken place as the lower part of the limb rotated down to contact the ground. The fore-limbs of the early amphibians display the same structure as those of all modern land-living tetrapods: there is a single bone (the humerus) in the upper arm, a pair of bones (the radius and the ulna) in the forearm, and a set of smaller bones composing the wrist and hand region.

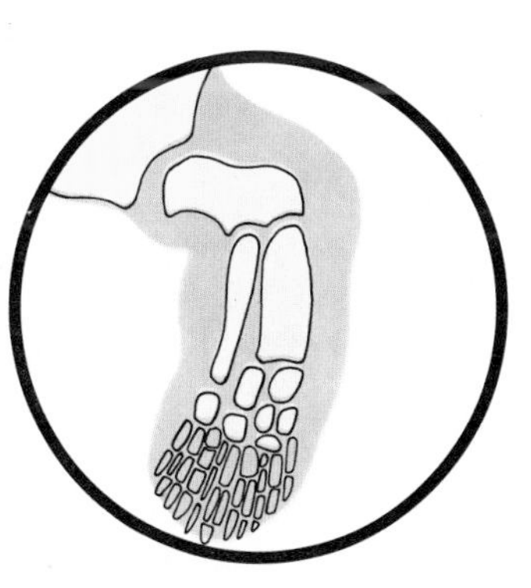
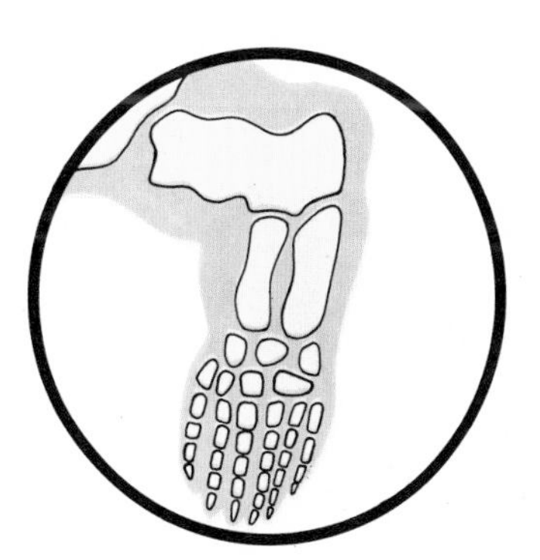
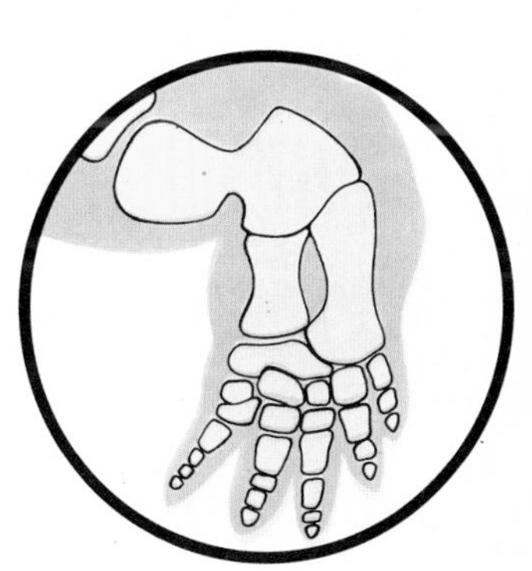

Like Fish Out of Water?
Although amphibians remained fish-like for many millions of years, Devonian and Carboniferous fossils show that their bodies underwent great changes. The bones in their legs became thick enough to support their body weight, and the leg muscles powerful enough for them to crawl or walk. The backbone also became sturdier, to support the weight of the body hanging below it. Eardrums evolved which were sensitive enough to detect airborne sounds, which are much fainter than sounds travelling through water. The eyes developed eyelids for protection and were kept moist and clean by tear glands.

Above: Most modern amphibians pass through an aquatic tadpole stage. The frog tadpole (left) is just losing its tadpole tail and taking on the adult shape. A sequence of branchiosaur fossils (centre) discovered in Carboniferous rocks shows that the early amphibians had a similar life history. Some modern amphibians, such as the axolotl (right), never grow up completely and they retain their gills all their lives.

Below: Some Carboniferous lepospondyls: Phlegethontia (left), Ophiderpeton (centre) and Microbrachis.

Sham Conquerors
It is often said that the amphibians conquered the land. The truth is, perhaps, a little less impressive. By some 350 million years ago, amphibians had become able to spend all or part of their time on land. But most of them remained dependent on the water for breeding and feeding.

Even today most amphibians start their lives in the water. The eggs are usually laid in water, and hatch into fish-like larvae with feathery gills. These larvae, or tadpoles, feed in the water and, after some weeks or months, they lose their gills and develop lungs as they 'change' into adults.

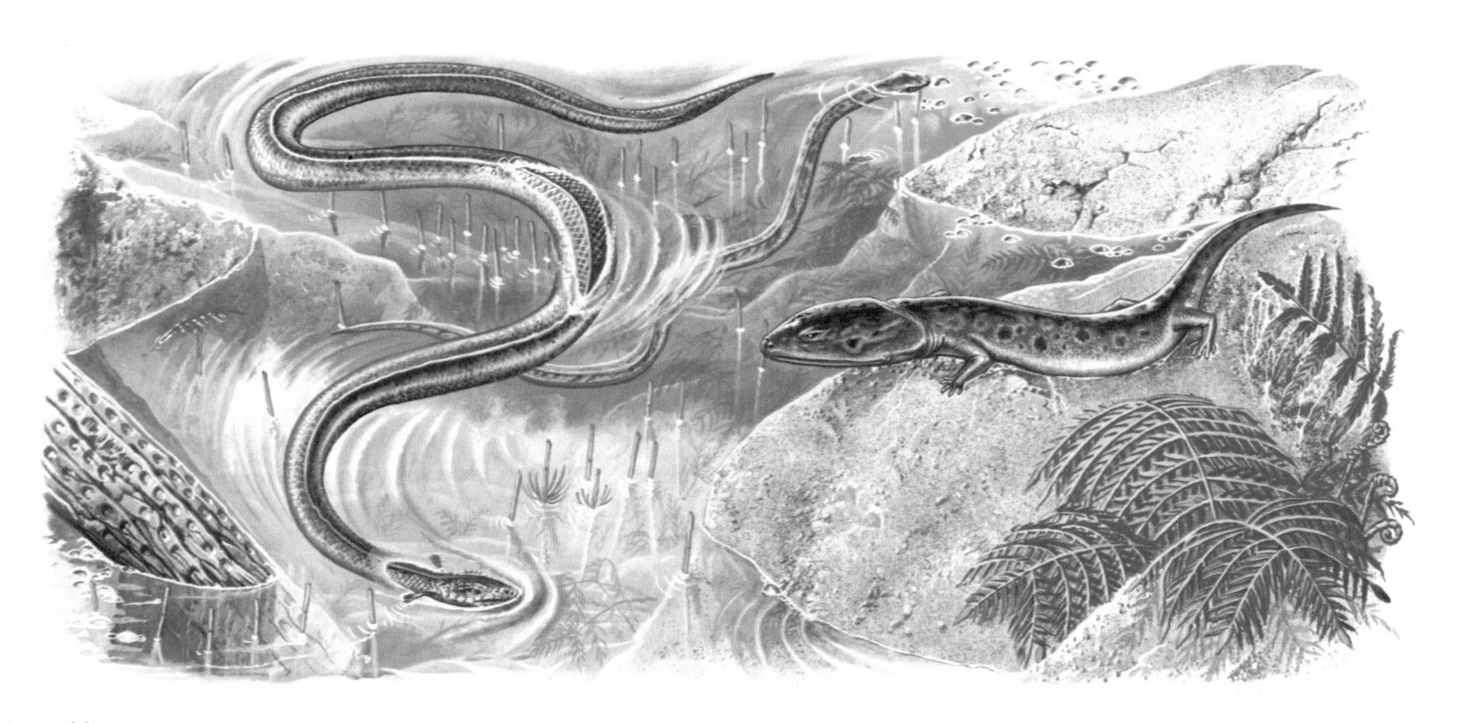

The Swamp Dwellers

The earliest known amphibian is *Ichthyostega*. Like its fishy ancestors, it had a long, finned tail and was an excellent swimmer. It probably spent most of its time in the water. *Ichthyostega* was about one metre (three feet) long and belonged to the largest group of amphibians, the labyrinthodonts.

Another group of amphibians was that of the lepospondyls. These skinny creatures lived in late Carboniferous times. Many of them must have spent all their time in the water. Some, such as *Ophiderpeton* and *Phlegethontia* had even lost their redundant limbs. They swam about in the swampy pools by moving their tails from side to side, like aquatic snakes. Other lepospondyls, such as *Microbrachis,* retained small limbs and were able to leave the water, like newts. *Diplocaulus* had limbs, too, but it probably spent most of its time on the bottom of lakes, preying on tiny water creatures and the larvae of other amphibians. A few Permian lepospondyls retained stout limbs, however, and probably spent most of their time on land, like living salamanders. It seems likely that the ancestors of modern amphibians are to be found among this group.

TERTIARY
CRETACEOUS
JURASSIC
TRIASSIC
PERMIAN
CARBONIFEROUS
DEVONIAN
Frogs And Toads
Newts And Salamanders
Apodans
To Reptiles
Labyrinthodonts
Lepospondyls
Lobe-finned Fishes

Above: The amphibian stock divided early on into two main lines – the labyrinthodonts and the lepospondyls. Either of these may have been the ancestors of modern amphibians. The reptiles probably arose from the labyrinthodonts. Many smaller groups of amphibians arose and became extinct on the way.

No Enterprise – or No Evidence?

Many explanations have been suggested for the timidity of the Carboniferous amphibians' so-called conquest of the land. Even in the Permian Period most of the labyrinthodonts were still aquatic. This may have been partly because the early amphibians were carnivores, used to a diet of fish, and there was not yet enough for them to eat on land. Or it may have been because the reptiles, which evolved soon after the amphibians, outstripped them in the competition for food and space on land.

It is also possible that many more

Above: The larger Carboniferous amphibians belonged to a group called the labyrinthodonts. Some of these – the anthracosaurs – were aquatic creatures like Eogyrinus, which has been found in coal deposits in England. Its large tail and elongated body suggest that it swam by undulating its body from side to side. It probably lived in lakes and streams and fed on fish. Similar fossils have been found in North America.

Right: Diplocaulus, a lepospondyl from the Permian, seems to have been a pond-dweller. The 'Napoleonic' head-shield may have been a form of defence: it would certainly have been very difficult for other creatures to swallow.

THE FLOOD WATCHER

In the days before the theory of evolution had been put forward the presence of fossil skeletons in the rocks needed some explanation. One explanation was that these were the remains of animals drowned in the biblical flood. A leading supporter of this idea was the Swiss, Johannes Scheuchzer. In 1726 he published a description of a skeleton from Germany which he believed to be that of 'Homo diluvii testis' – a man who had witnessed the flood. Since the skeleton was only 60 centimetres (two feet) long, it would have been a rather small man, and we now know that it is really the skeleton of a giant salamander, about 8 million years old. It has been named *Andrias scheuchzeri.*

Scheuchzer also examined a number of fossil plants. Because he thought that many of the fossil seed cones were in the condition one would expect in the spring, he stated that the 'flood' must have taken place in May!

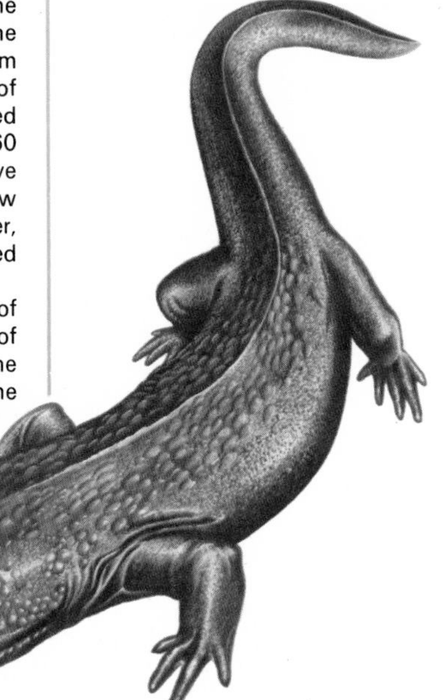

Andrias scheuchzeri – not a man but a large salamander.

amphibians did manage to survive on land and that we know little of them simply because they are absent from the fossil record. Almost all our knowledge of Carboniferous amphibians comes from fossils preserved in the coal deposits – the remains of lush swamp vegetation. It is not surprising to find that nearly all the animals that lived in these forests were adapted to an aquatic existence. Unfortunately, we do not know much about the animals that might well have been living on the higher, drier ground above the swamps.

The Permian Amphibians

About 280 million years ago, parts of Texas were covered by a great river delta. The sediments deposited in this delta contain the remains of many amphibians and reptiles, giving us our next glimpse of early land life. The early Permian amphibians included both lepospondyls and labyrinthodonts. Unlike today's amphibians, they had dry skin, often with scales or bony plates embedded in it. This armour presumably gave them some protection against the reptiles.

Unlike any of their predecessors, some of the Permian amphibians appear to have been completely terrestrial. *Cacops* had slender limbs and probably spent almost all its adult life on land, feeding on the ever-growing numbers of insects and other invertebrates. It was about 40 centimetres (16 inches) long. Another small terrestrial

Below: A scene from the Texas delta during the early Permian Period. Dominating the picture is the giant labyrinthodont Eryops (foreground right and background left). Eryops spent most of its time in the water hunting for fish. The trio on the sun-baked red rocks behind are Diadectes and two seymouriamorphs. In the centre of the picture are two Cacops. Well armoured, they seem to have been almost completely terrestrial. The tall trees in the background are conifers called Walchia. The foliage in the foreground, left, is Gigantopteris.

form was *Seymouria* which grew to a length of about 60 centimetres (two feet). A relative of *Seymouria,* called *Diadectes,* was especially well-adapted for life on land. It was up to three metres (ten feet) long, and its flattened, grinding teeth suggest that it was one of the first land-living herbivores.

Even in the Permian Period, however, most amphibians seem to have been at least partially aquatic. *Eryops*, a lumbering creature nearly two metres (six feet) long, must have led the same sort of life as a modern crocodile. With its cumbersome body and stubby legs, it must have been a slow and clumsy mover on land. It probably lazed in the shallows snapping up other creatures with its powerful jaws. In the deeper water was *Archeria,* a relative of the Carboniferous *Eogyrinus*. With its slender body and tail fin, it must have been a good swimmer. Assured of a constant supply of fishes and young amphibians to eat, it probably left the water only rarely.

Farewell to the Land

By late Permian times, the labyrinthodonts had become extinct on land and the 'invaders' were beating a retreat to the water. The reptiles were becoming progressively more varied and active, and the amphibians could not survive their competition. A few of the larger, heavier types survived in the water until late in the Triassic, leading a sort of sluggish, crocodilian life. Other more highly specialized amphibians, such as *Aphaneramma* and *Gerrothorax,* survived alongside them, but soon they too declined, leaving only a handful of representatives in the modern world. These – the frogs and toads, the newts and salamanders, and the limbless apodans – are rather specialized descendants of the Palaeozoic amphibians. Most of them still live in or near water.

Above: These two Triassic amphibians were both aquatic, but their life styles were very different. Aphaneramma (left) had a skull and snout up to 20 centimetres (eight inches) long and it actively hunted fish in the sea. Gerrothorax was about one metre (three feet) long and it had a very wide, flattened head. Like today's angler fish, it may have lurked in deep water, attracting prey by means of a brightly-coloured lure inside its mouth.

AMPHIBIAN OR REPTILE?

The dry skin, large size, and terrestrial habits of some of the early Permian amphibians make them appear very much like reptiles. It is not surprising that some of them were once thought to be early reptiles. The essential difference between the amphibians and the reptiles lies in the reptiles' ability to lay shelled eggs. Unfortunately, this feature does not help scientists to decide whether a fossil skull belonged to a late amphibian or an early reptile, and even now it is not always possible to say which group some of the Permian fossils belong to. But the skulls do sometimes hold clues which may solve the problem. For example, young specimens of the terrestrial *Seymouria* still bear on their skulls traces of a system of grooves which were used to detect sound waves in the water. *Seymouria,* therefore, clearly went through an aquatic stage during its early life, and was thus an amphibian. *Hylonomus* (top) was one of the earliest reptiles.

Dimetrodon was a pelycosaur which lived in Permian Texas. With its sharp cutting and stabbing teeth, it must have been the dominant predator. It probably ate other reptiles as well as amphibians. Like nearly all reptiles, it must have laid eggs. It may have guarded its nest as the crocodiles do today but it would probably normally have abandoned its young soon after they hatched. Because small animals have a high surface area in relation to their volume, the young probably did not need a high sail to regulate their body temperature.

The Rise of the Reptiles

It was probably quite early in the Carboniferous that the reptiles evolved from the amphibians. Destined to rule the world for nearly 200 million years, they lost no time in colonizing the land. The key to their initial success was the ability to lay a hard-shelled egg which could survive on land and provide nourishment for the growing embryo. This set the reptiles free from the watery chains that had shackled the amphibians.

Although the development of an egg with a tough shell must have given the reptiles a considerable advantage over the amphibians, their bodies must have been superior in other ways, too. Otherwise it is difficult to explain why the reptiles quickly became so much larger than any terrestrial amphibian. Changes in their skeletons and muscles allowed the reptiles

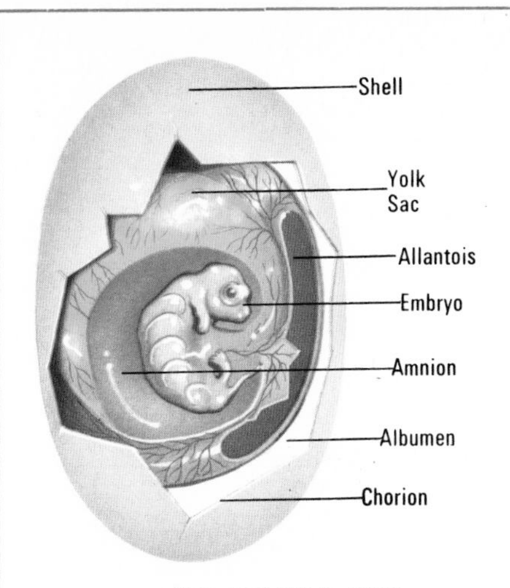

THE REPTILE EGG

Eggs look simple objects, but beneath the hard or leathery shell lie various membranes that are vital to the developing embryo within. The shell is vital, too, for it both supports and contains the weight of the embryo and its reserves of food and water. It also reduces the rate of evaporation of water.

Essential as it is, the shell also presents problems. It hinders the exchange of oxygen and carbon dioxide between the embryo and the air. To overcome this, a special membrane called the chorion lies just under the shell and bears a network of blood vessels which transport these gases to and fro. There is also a danger that the embryo might become pressed against the shell and deformed, but this danger is avoided because the embryo lies within another liquid-filled membrane called the amnion.

The embryo must also absorb food and excrete waste. Blood vessels in the membrane around the yolk carry dissolved food to the embryo, and the excretory products are deposited in another membrane called the allantois. The chorion, amnion, and allantois are all left behind when the embryo hatches.

Above: The bulky body and small head of Edaphosaurus suggest that it was a herbivore. It had a powerful battery of crushing teeth which would have been well suited to grinding up the tough leaves of the abundant seed ferns. The animal may also have used its teeth to crush and eat mollusc shells. Unlike that of Dimetrodon, the sail must have been quite thick, for the bones supporting it carried a series of small cross bars, rather like those of the mast of a sailing ship.

THERMOSTAT SAILS

Reptiles lack an insulating layer of hair or feathers, and their body temperatures rise and fall with that of the surroundings. Many living reptiles can regulate their temperatures to some extent, however, by lying broad-side to the sun if they are cold, especially in the morning, and by retreating into the shade if they are too hot. The sails of the pelycosaur reptiles may have aided in this type of behaviour by providing a larger surface area over which heat could be absorbed or radiated. If this were so, we would expect the sails to be proportionately larger in the bulkier animals and proportionately smaller in the more slender creatures. This is exactly what we find in the fossils.

Left: The oldest known reptile egg was found in early Permian rocks in Texas. There is, of course, no way of discovering exactly which reptile laid it, but it is nearly six centimetres (two and a half inches) long and must therefore belong to one of the larger mammal-like reptiles such as Dimetrodon or Edaphosaurus. X-ray photographs have shown no trace of an embryo, and, since the egg clearly did not hatch, it was probably addled.

to move more efficiently on land, and their skin became thicker and more waterproof. Perhaps their breathing mechanism, or their heart and blood systems were also more effective than those of the amphibians. They may also have begun to develop the ability to keep their body temperature constant.

The first reptiles, the cotylosaurs, looked like amphibians and are usually known as *stem reptiles* because they gave rise to all the others. By the beginning of the Permian Period the land was already dominated by reptiles – not by the mighty dinosaurs, but by the smaller mammal-like reptiles. Though these creatures were less spectacular than the dinosaurs, they were in other respects more important, for they eventually gave rise to the mammals.

Below: Titanosuchus (left) and Moschops (right) were two of the large mammal-like reptiles in the Karroo Community of South Africa in Permian times.

The Sail Backs

The earliest mammal-like reptiles were the pelycosaurs. Several different types are known, all with a clumsy, sprawling posture. Some, such as *Dimetrodon* and *Edaphosaurus,* wore an enormous 'sail' on their backs, which may have helped them to keep their body temperature constant (see panel on page 93).

Many of the pelycosaurs lived in swampy areas and fed on a diet of fish, shellfish, and small land animals. *Dimetrodon* and *Ophiacodon*, a sail-less form, were both carnivores. Among some of the other pelycosaurs, however, a brand new way of life was emerging, for they were beginning to eat plants. This habit required several novel adaptations. The animals had to evolve teeth which could cope with the continual wear caused by the hard cell walls of the plants. Plant material also takes longer to digest than meat, and the herbivores needed bulkier bodies in which to store their food. Some grew very large indeed. *Cotylorhynchus,* which belonged to a group of herbivorous pelycosaurs called the caseids, was three metres (ten feet) long and weighed up to 330 kilograms (52 stone).

Below: In this scene from the late Permian Karroo, the gorgonopsian carnivore Lycaenops (right) is seen attacking a dicynodont. These vegetarians were the commonest vertebrates in the Karroo fauna at this period. Two more predators (centre) lurk near the scene of the kill in hopes of a meal. They belong to the therocephalian order, and are related to Lycaenops but show further developments towards a mammalian physiology. In the background (left), a group of pareiasaurs are grazing peacefully on the fringes of a swamp. Their coarse, bone-studded hide gave them some protection against predators. The small insect-eater in the foreground (right) is Galechirus.

The Karroo Community

The early Permian reptiles were, as far as we know, confined to the old supercontinent of Euramerica, but the story now switches to Africa – in particular to the region in the south known as the Karroo. The Upper Permian rocks of that region are exceptionally rich in vertebrate fossils. Similar fossils are also known from India and Russia, and it is clear that by this time land vertebrates had spread right across the world.

Beaks and Daggers

In the Karroo fauna, we see for the first time a community not very unlike the animal communities we know today. There were many types of bulky herbivore, each adapted to a particular kind of plant diet, and a more limited variety of smaller carnivores. The fossils of herbivores and carnivores can be distinguished in several ways. The herbivores have smaller heads, blunter teeth and bulkier bodies, and they are always much more numerous than the carnivores which prey on them.

About 80 per cent of the Karroo fossils belong to a group of mammal-like reptiles called dicynodonts. These were all herbivores. Many of them had a horny beak

which was continually replaced as its surface was worn away by the constant cutting and grinding of plant material. The animals probably browsed in herds, keeping a watchful eye open for the predators that hunted them. Among these were gorgonopsids, such as *Lycaenops,* whose gaping mouths revealed dagger-like teeth. With their slender bodies and legs they could outrun their clumsy prey and pierce the plant-eaters' thick hide with a single stab of the teeth.

The Mammals Wait

By late Permian times, an even more successful group of mammal-like reptiles had emerged. These were the cynodonts, the group that later gave rise to the mammals. The earlier cynodonts lived alongside the dicynodonts and gorgonopsids in Africa. They were slender, agile creatures, about the size of a cat. With teeth like those of a modern shrew, they were probably omnivores or insect-eaters. Their bodies had many mammalian features, one of which may have been a scanty covering of hair. Although these hairs would not at first have formed an effective insulating layer, they represented the beginning of the mammalian method of regulating temperature.

As the first mammals emerged, their ancestors began to decline. By the end of the Triassic Period the mammal-like reptiles were all extinct. But the mammals could not claim their inheritance immediately. The dominion of the land was already held by another group of reptiles – the dinosaurs. Before continuing with the story of the mammals we must look at the 'terrible lizards' that overshadowed them for so long.

Robert Broom (1866–1951)

ROBERT BROOM

Robert Broom, one of the most brilliant investigators of the evolutionary history of mammals, was born in Scotland in 1866. He went to Edinburgh University and became a doctor. His interest in mammalian evolution then took him to Australia, where he studied the primitive monotremes and the marsupials. After seeing fossils of mammal-like reptiles from South Africa, Broom settled in that country in 1897 and he eventually became a professor of zoology and geology at Stellenbosch University. By 1908 he had revolutionised our understanding of the mammal-like reptiles and he had shown that the South African forms had descended from the pelycosaurs of North America.

But perhaps the most remarkable part of Broom's career was yet to come. The first African 'ape-man', *Australopithecus,* was discovered in 1924 and Professor Raymond Dart (see page 149) decided that it was a close relative of man. Broom supported Dart in this belief and, in 1936, he began to search for *Australopithecus* himself. Almost immediately he found a nearly complete skull, and he unearthed several more specimens later. Although opposed by anthropologists in Europe and North America, he suggested and then convincingly proved the evolutionary relationships between the australopithecines and man. This new work, which completely altered our understanding of man's early evolution, did not begin until Broom was already 70 years old. It continued until his death in 1951.

The Mesozoic World

The Mesozoic Era is divided into three periods – the Triassic, the Jurassic and the Cretaceous. It lasted for about 160 million years. This vast span of time witnessed great changes in the faunas and floras of the earth. Continents broke up and climates altered; the reptiles took over land, seas and skies; and mammals made their first tentative appearance.

Patterns of Land

Not long before the Mesozoic began, the land masses of the world had joined up into a single supercontinent, Pangaea. In the Carboniferous and early Permian Periods large areas around the South Pole had been glaciated, but now the ice had disappeared. Though mountains and deserts existed, they did not isolate any individual land area. For the first time, terrestrial animals were able to spread throughout the world. This remained true in the Triassic Period, and as a result identical types of reptile are found in every one of the regions which were destined to become the separate continents of the modern world.

The Triassic land mass did not long remain a single continent. Early in the Jurassic the first split appeared. The old southern land mass, Gondwanaland, began to edge away from 'Euramerica' and a narrow but deep sea formed between the two continents. They must still have been linked by a land bridge, however, for similar types of dinosaur are found in North America and East Africa. Elsewhere shallower seas advanced across the land. By the middle of the Jurassic, a wide but shallow stretch of water, the Turgai Sea, separated Asia from Europe.

By the early Cretaceous Period, the Atlantic Ocean had widened, and the northern and southern land masses had separated completely. A tongue of sea had spread around the eastern and southern edges of what was to become the African continent.

Oceans and seas became even more widespread in the later part of the Cretaceous. South America and Africa were now isolated. Shallow seas crossed both North America and Eurasia from north to south, creating a barrier between the

Below: In the Triassic Period, all the land areas of the world were united into one great land mass, known as Pangaea.
In the Jurassic, the southern land mass, known as Gondwanaland, began to move away from the northern land mass, Laurasia, but a land bridge still connected the two supercontinents.
Shallow seas (light blue) started to spread across them. In the early Cretaceous, Gondwanaland and Laurasia became totally separate.
In the late Cretaceous, South America, Africa and India became distinct. North America remained united to Eurasia, but shallow seas bisected the land area of North America and separated Europe from Asia.

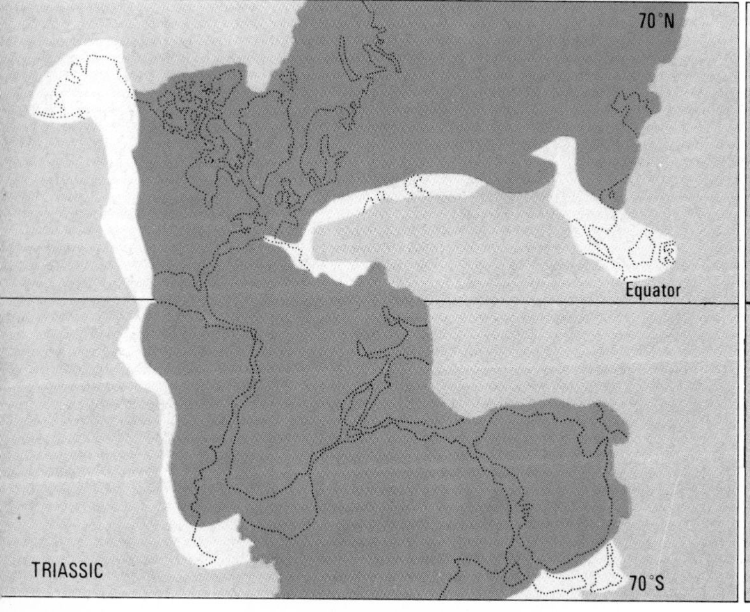

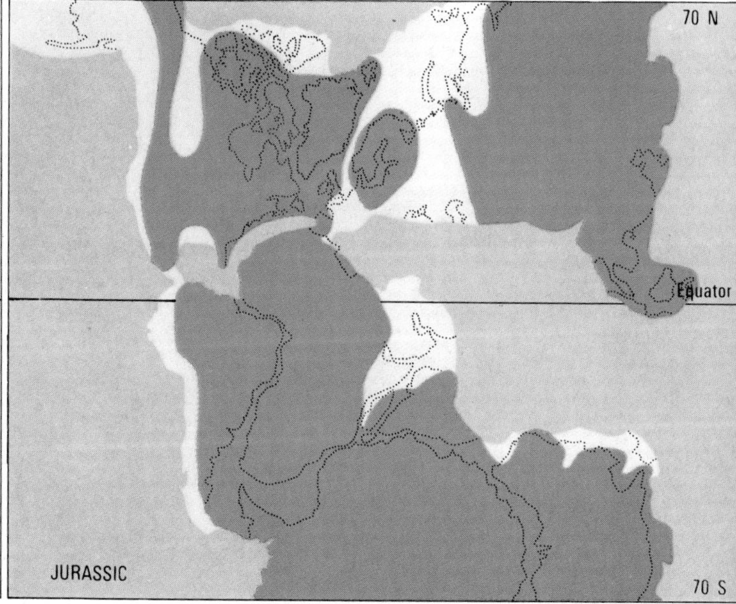

western land mass (consisting of Asia and western North America) and the eastern land mass (consisting of eastern North America and Europe). This meant that the new types of dinosaur which evolved at this time in the western continent were unable to spread to Europe, where a more old-fashioned dinosaur fauna lingered on.

Changing Climates

It was not only the geography of the Mesozoic world which was affected by continental drift. Climates altered too. Large land areas that lie far from the sea, like Central Asia today, have a much harsher climate – with greater seasonal extremes – than areas closer to the sea. The spreading oceans and seas of the Mesozoic thus brought milder climates to most of the world; the absence of polar ice caps during this period must also have had a warming effect. Mild climates which suited both reptiles and land plants now extended as far north as Alaska and Siberia.

The creation of mountain ranges, such as the Andes and the Rockies, also affected Mesozoic climates. As rain-bearing winds approach these lofty barriers, they are forced to rise and drop their heavy load of water. The windward slopes thus experience a wet climate, while the lee slopes are in a permanent rain shadow.

Above: The Cretaceous Period gets its name from the Latin word 'creta' meaning chalk, because very thick beds of chalk (left) were formed from the shells of the tiny marine organisms that flourished in the warm seas of those times. The delicate cart-wheel structure of one of these organisms, a coccolith, can be seen in the photomicrograph (right). It has been magnified more than 10,000 times.

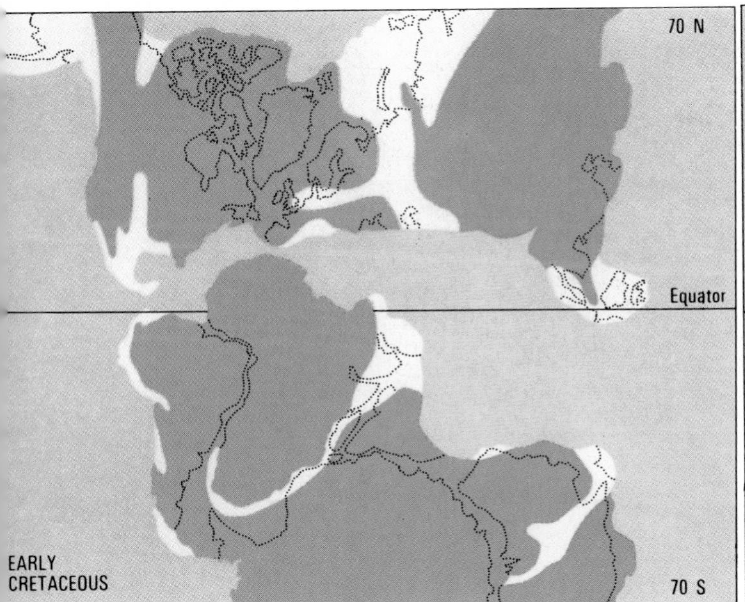

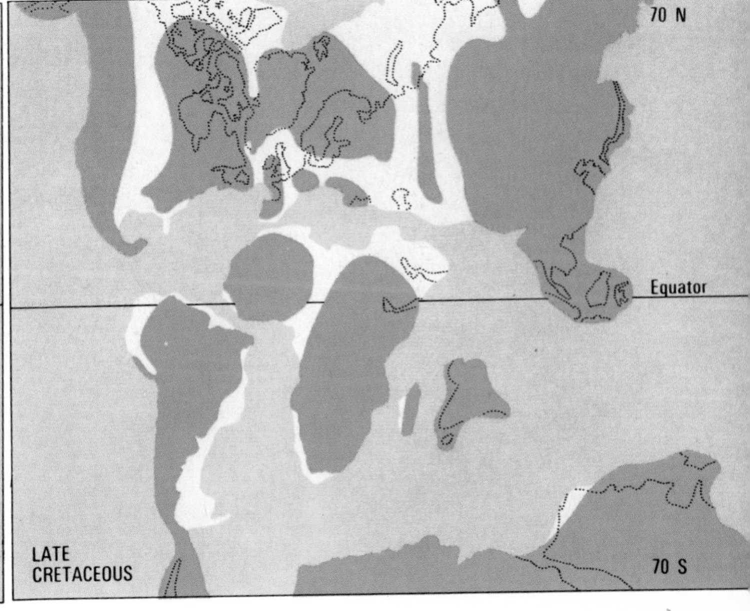

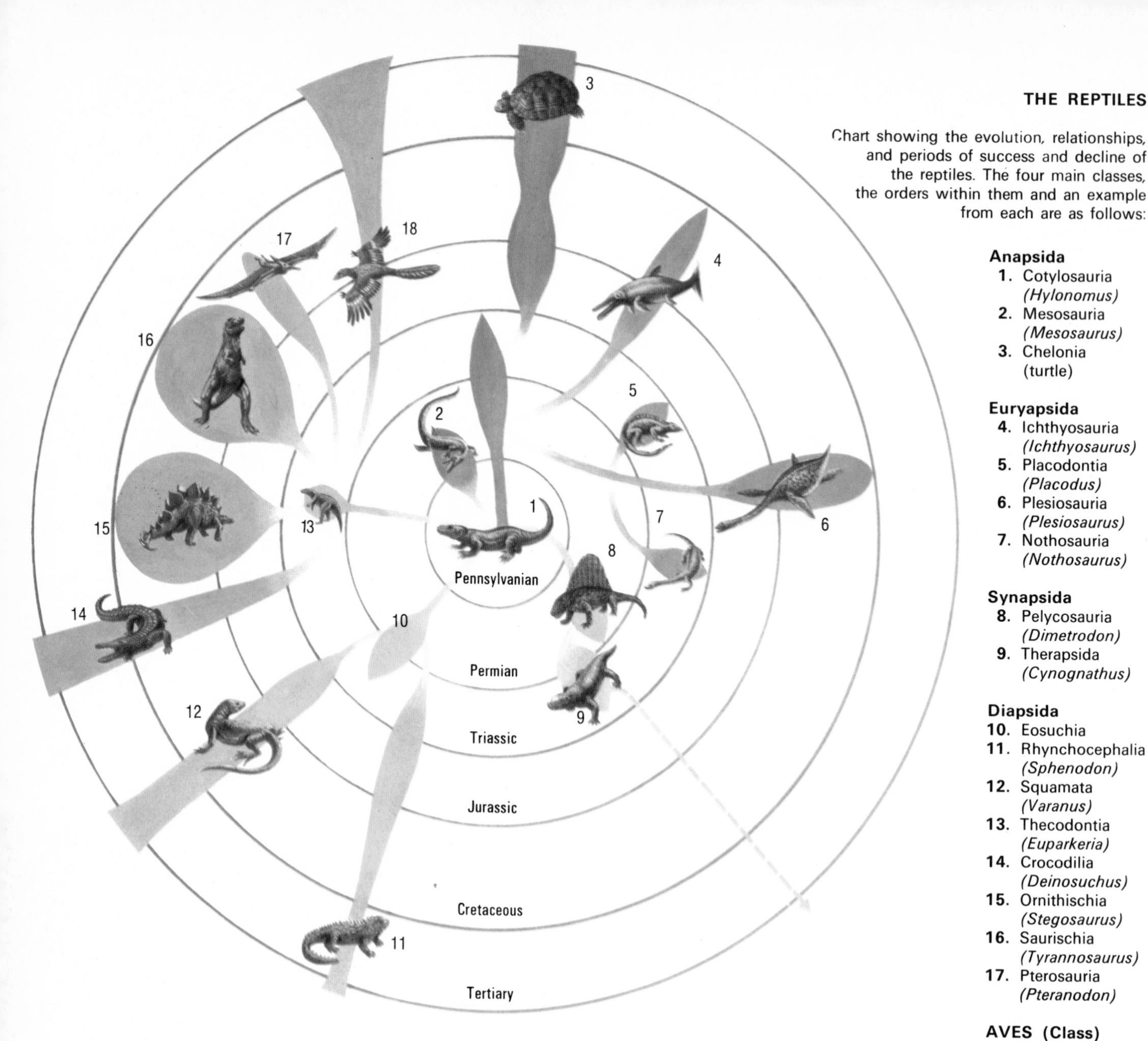

THE REPTILES

Chart showing the evolution, relationships, and periods of success and decline of the reptiles. The four main classes, the orders within them and an example from each are as follows:

Anapsida
1. Cotylosauria (*Hylonomus*)
2. Mesosauria (*Mesosaurus*)
3. Chelonia (turtle)

Euryapsida
4. Ichthyosauria (*Ichthyosaurus*)
5. Placodontia (*Placodus*)
6. Plesiosauria (*Plesiosaurus*)
7. Nothosauria (*Nothosaurus*)

Synapsida
8. Pelycosauria (*Dimetrodon*)
9. Therapsida (*Cynognathus*)

Diapsida
10. Eosuchia
11. Rhynchocephalia (*Sphenodon*)
12. Squamata (*Varanus*)
13. Thecodontia (*Euparkeria*)
14. Crocodilia (*Deinosuchus*)
15. Ornithischia (*Stegosaurus*)
16. Saurischia (*Tyrannosaurus*)
17. Pterosauria (*Pteranodon*)

AVES (Class)
18. *Archaeopteryx*, the first known bird.

The variety of climates produced as a result of these climatic conditions afforded a great deal of scope for the evolution of Mesozoic plants and animals.

The Age of Reptiles

Against this background of change, a great explosion of reptilian evolution took place. Aptly named the Age of Reptiles, the Mesozoic Era saw the land dominated by dinosaurs, the skies by pterosaurs, and the seas by their aquatic relatives.

On land, the plants too were changing. The first flowering plants, or angiosperms, appeared early in the Cretaceous, and by the middle of this period they had spread all over the world, gradually ousting the gymnosperms from their position of supremacy.

Below: A crocodile. The crocodilians probably escaped extinction at the end of the Mesozoic Era because they lived in the fresh waters. They are the only living archosaurs.

Flowering plants still dominate the land floras today. But the Mesozoic reptiles have all disappeared. Quite suddenly, at the end of the Cretaceous, they died out and the mammals took over as the dominant land animals. Perhaps the uninsulated reptiles were unable to survive the climatic changes which were taking place at this time.

In the seas, too, there were widespread extinctions. The ammonites and most of the belemnites disappeared for ever. It has been suggested that marine food chains were disrupted at the end of the Cretaceous by a reduction in the amount of mineral matter washed down from the land by erosion. This would create an unfavourable environment for the planktonic plants on which all marine animals depend.

These abrupt changes make it easy for palaeontologists to define the boundary between the Mesozoic and Cenozoic Eras, for each possessed a completely distinctive fossil fauna.

Right: A fossil leaf from a Cretaceous plane tree in Greenland. The flowering plants began their march to dominance early in the Cretaceous and, by the end of the period, had spread to all parts of the world.

Below: The Andes started to rise in the late Mesozoic when South America began to drift westwards (see page 55).

The Rule of the Reptiles

Tanystropheus lived around the shores of the Muschelkalk Sea in Triassic Europe. The adult animal appears to have fed on fish and cephalopods. Puzzlingly, fossils of the younger, smaller individuals have not been found in the same marine sediments but in deposits formed closer to land. It has recently been suggested that the young Tanystropheus spent its early life on land and used its long neck to catch flying insects!

When one thinks of the Age of Reptiles, one naturally thinks first of the dinosaurs. Yet these animals formed only a part of the great radiation of reptiles that took place during the Mesozoic Era.

Palaeontologists divide the reptiles into four groups – the synapsids, diapsids, anapsids and euryapsids (see chart page 98). These distinctions are based on differences in the skulls. The synapsids include the pelycosaurs and the mammal-like reptiles. Although these early reptiles gave rise to the mammals, they did not survive long beyond the beginning of the Mesozoic. The diapsid group contains the snakes and lizards and the tuatara (see panel), together with all the 'ruling reptiles', or archosaurs. These include the crocodiles, the ancestors of the birds, the pterosaurs, and the two groups of dinosaurs – the Saurischia and the Ornithischia.

Of the remaining two groups, the anapsids mostly died out in the Palaeozoic. Only the clumsy tortoises and turtles, whose ancestors evolved during the Triassic Period, survive today. The last group, the euryapsids, contains aquatic reptiles such as the ichthyosaurs and plesiosaurs.

Rhynchosaurs have been found in Middle and Upper Triassic beds from almost all parts of the world except Australia. Scaphonyx (below) was a weighty animal from Brazil. Its bulk and nutcracker jaws strongly suggest that it was a herbivore. Despite their abundance, the rhynchosaurs did not survive for long. By the end of the Triassic, they had all disappeared.

Changing Fauna

On land, major changes in the earth's reptile fauna took place during the Triassic Period, when the ancestors of the dinosaurs displaced the last mammal-like reptiles. The reasons for these changes are still uncertain. It is strange that the mammal-like reptiles should have been sufficiently advanced to produce the earliest mammal by the end of the Triassic, but that neither they nor the mammals were able to compete successfully with the archosaurian reptiles. Perhaps they simply could not move about as efficiently as the reptiles, or perhaps they were not so skilful at producing and rearing their young.

A Brief Appearance

As the mammal-like reptiles (synapsids) declined, but before the full array of their dinosaur successors appeared, a few other reptiles achieved a brief period of success. One group in particular – the rhynchosaurs – became quite abundant for a time. Distantly related to the tuatara, the rhynchosaurs were rather heavily-built animals, about two metres (six feet) long. Their toothless jaws were like a massive pair of bony tongs, operated by enormous muscles. Inside the mouth they had huge, crushing tooth-plates. Some scientists think that these jaws may have been an adaptation for feeding on molluscs; the animals could have dug up the shells and grasped them with their beaks, and then cracked them open with their tooth-plates. Rhynchosaurs were so abundant in some areas, however, that it seems more likely they were herbivores. They probably fed on fruit, using their beaks to crack nuts and rip off the husks, like a parrot.

Snakes and Lizards

The most successful of all the reptiles are the group known as the Squamata – the lizards and snakes. The lizards evolved at the beginning of the Triassic and many species are still with us today. Modern lizards include the iguanas, geckos and chameleons.

From its earliest days, the group contained some remarkably specialized animals. Two of these were *Kuehneosaurus* and *Icarosaurus*, both of which – like the modern lizard *Draco* – had elongated ribs projecting from their flanks. In *Draco*, these ribs are covered with a membrane of skin, and allow the animals to glide for distances of up to 13 metres (45 feet) with little loss of height. *Kuehneosaurus* and *Icarosaurus* may have been able to glide in the same way.

A weird creature called *Tanystropheus*, which lived in Europe in Middle Triassic times, was probably an early offshoot of the lizards. Though the total length of the animal was up to six metres (20 feet), the body and tail accounted for only 10 and 20 per cent of this respectively. The remainder consisted of a small head carried on the end of an immensely elongated neck. This neck was made up of only 11 vertebrae, some of which were 25 centimetres (nine inches) long.

Tanystropheus seems to have lived in and around the shallow seas which at that time covered central Europe. Fossilized stomach contents show that it fed on fish and cephalopods, which it probably caught by making sudden darts with its long, flexible neck.

After their initial flourish, the lizards faded into comparative insignificance. One aquatic group, the mosasaurs (see page 104), became quite important in the Cretaceous Period, but most lizards lived on under the shadow of the dinosaurs and

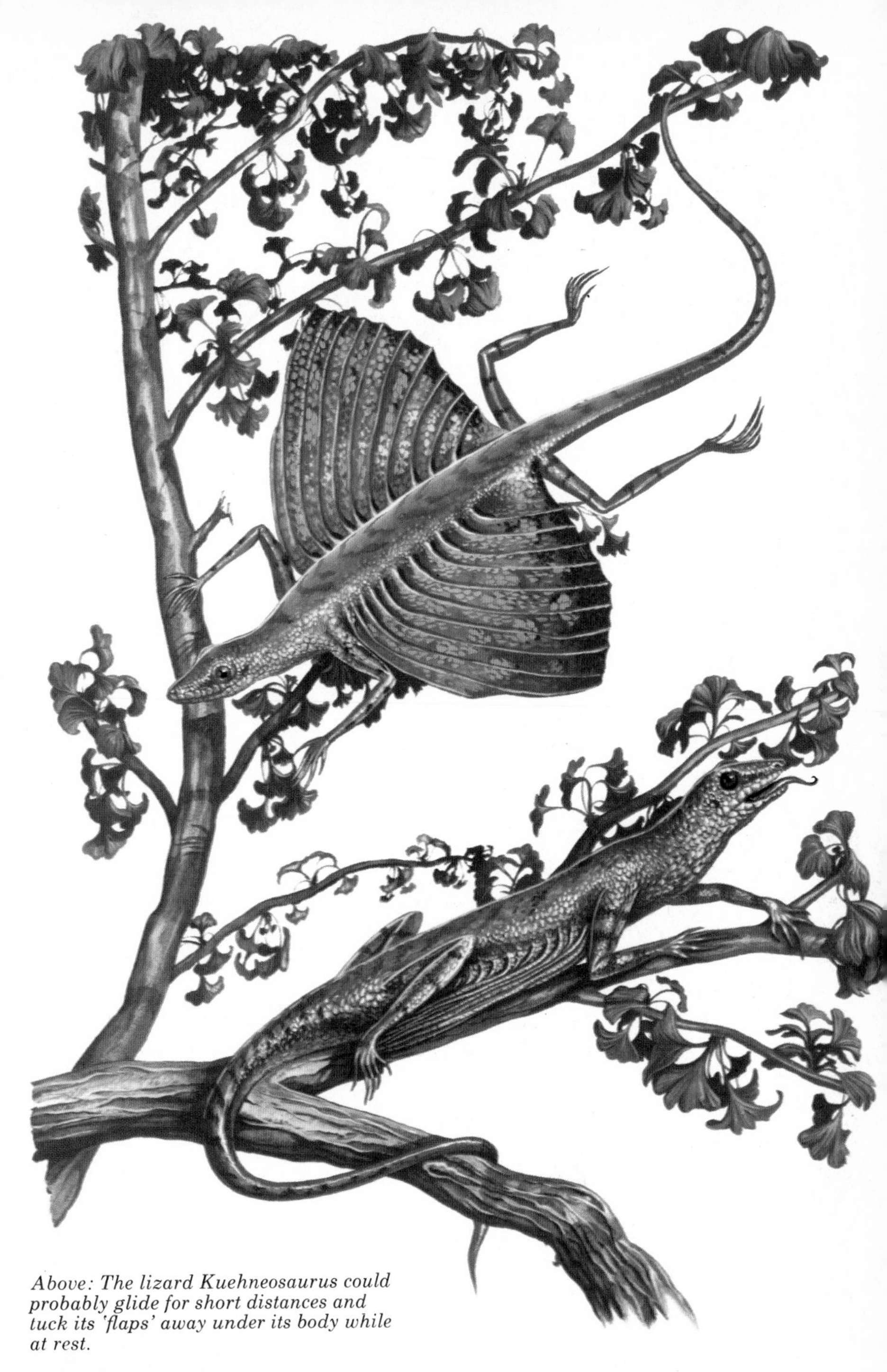

Above: The lizard Kuehneosaurus could probably glide for short distances and tuck its 'flaps' away under its body while at rest.

THE TUATARA

The tuatara is the only surviving rhynchocephalian. A living fossil, it survives today only in the protected isolation of about 20 small islands off the coast of New Zealand. Up to 60 centimetres (two feet) long, the tuatara lives in burrows by day and is active at night. It feeds on invertebrates, and occasionally on the eggs or young of petrels. The petrels, too, make burrows, and tuatara and petrels are sometimes found living together in a shared burrow. Tuataras have a slow metabolic rate and are very slow-growing. Studies suggest that they take 20 years to reach maturity, continue to grow for 50 years and may live for a century or more.

The pliosaurs were voracious, short-necked carnivores. Kronosaurus (above) lived in the Cretaceous seas and could grasp even large reptiles in its powerful jaws. Peloneustes (below) was also a fish-eater. Though its whole body was less than the length of Kronosaurus's head, it was well equipped as a hunter, with sharp teeth to grip its slippery victims.

into the modern world only as small, shy carnivores. Their limbless relatives, the snakes, appeared at the end of the Cretaceous when the dinosaurs were dying out. But in a world full of mammals they, too, have been forced to lead a sheltered, unobtrusive life.

Back to the Water

The reptiles were the first vertebrates to become completely independent of the water. A complex series of adaptations had enabled them to adjust to life on land. But no sooner had they attained this freedom than some of them returned to the water. In order to move, breathe, feed and reproduce in this strange, wet environment they had to undergo a reverse series of adaptations. The fishy characteristics which their ancestors had shaken off millions of years earlier had to be gradually reassumed. The versatile reptiles were undaunted: they not only adapted to the sea, but throughout the Mesozoic were absolute masters of the marine environment. The two most important groups, the plesiosaurs and the ichthyosaurs, both appeared in the late Triassic and survived until the late Cretaceous.

The Ichthyosaurs

Like modern dolphins, the ichthyosaurs were fast, powerful swimmers. They reached lengths of three metres (ten feet), and swam like fish by flexing their streamlined bodies from side to side. A powerful tail fin provided the main swimming thrust, and a fin on the back prevented the body rolling from side to side. The limbs were reduced to small paddles, used for steering and perhaps for braking.

Although the ichthyosaurs, like all reptiles, had to breathe air, they could not survive on land. Their small limbs could not support their great weight. For

this reason, they had to give birth to their young at sea. Some fossil specimens have been found in which the skeletons of small ichthyosaurs lie inside the adult, and sometimes even projecting from it, presumably through the birth passage.

Like the living marine mammals, the ichthyosaurs seem to have had a varied diet. With their long toothy jaws, they could hunt all kinds of fish, shell-fish and ammonites. Thousands of tiny hooks from ammonite tentacles have been found among the fossilized stomach contents of ichthyosaurs and other marine reptiles. One ammonite shell bears a pattern of tooth marks which accurately matches the pattern of teeth in the jaws of mosasaurs. (In similar fashion, the modern giant squid *Architeuthis* is pursued into the depths of the sea by the sperm whale.) One ichthyosaur, *Omphalosaurus*, seems to have been a mollusc-eater. Its jaws and teeth were ideally suited for crushing shells.

The Plesiosaurs

The true giants of the Mesozoic seas were the plesiosaurs. Some of these animals reached lengths of over 12 metres (40 feet). With squat bodies and stubby tails, they were slower swimmers than the ichthyosaurs. They propelled their huge bulk through the water by powerful strokes of their broad, paddle-shaped limbs. Studies of the skeleton suggest that the plesiosaurs could 'row' just as strongly backwards as forwards. They could probably turn swiftly by using a normal swimming stroke on one side of the body, while 'backing water' on the other side.

Some plesiosaurs had very long flexible necks. In one gigantic beast, *Elasmosaurus*, the neck contained 76 vertebrae and made up half the 13-metre (43-foot) length of the animal. This neck, lashing to and fro, must have struck terror into the fishes and smaller reptiles on which *Elasmosaurus* preyed. Another group of plesiosaurs, the pliosaurs, had short necks,

The ichthyosaurs seem to have occupied the same niche in ancient seas as dolphins today. Like dolphins, they were so well adapted to the water that they could swim as well as a fish and give birth to live young in the water. Some specimens of Ichthyosaurus (below) are so well preserved that the outline of the body can be clearly seen (above).

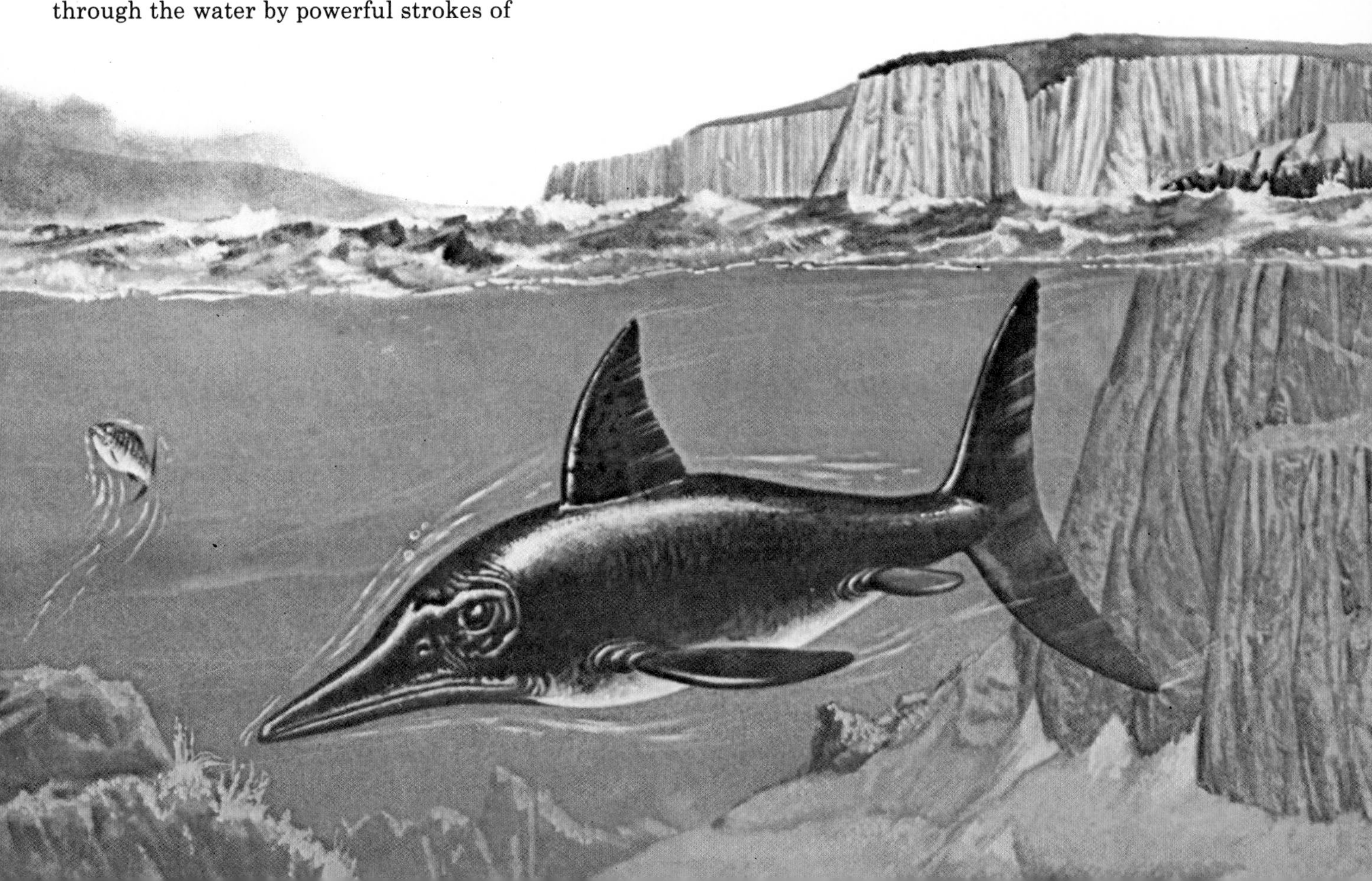

Above: The largest sea-turtle ever known was Archelon. Proof against predators, its shell was light enough for swift movement and hard enough to resist all but the most powerful teeth.

big paddles and huge heads. These included *Kronosaurus*, *Pliosaurus* and *Trinacomerium*. *Kronosaurus* was 17 metres (56 feet) long and its head measured four metres (13 feet). It probably preyed on other marine reptiles, just as killer-whales today feed on whales and seals.

Unlike the ichthyosaurs, with their smaller, weaker limbs, the plesiosaurs may have been able to scramble about on land. They may well have returned to the land to lay their eggs, as modern turtles do. A similar but smaller group of Triassic marine animals, the nothosaurs, almost certainly lazed on beaches like modern seals and sea lions.

Other marine reptiles of the Triassic Period included the placodonts and the mosasaurs. The placodonts probably lived in shallow water around European coasts: *Henodus*, *Placodus* and *Placochelys* belonged to this group. Their squat bodies were protected by heavy armour, and they used their strong beaks for gathering mussels from the sea floor and crushing them. The mosasaurs were marine lizards. *Tylosaurus*, a late Cretaceous mosasaur, reached more than 17 metres (56 feet) in length, and swam by means of its long tail.

Tortoises and Turtles

Like obsolete tanks, long forgotten by modern armies, the turtles and tortoises have lived on only as zoological curiosities. The earliest turtles were land-dwellers, and lived in the late Triassic. *Proganochelys*, like its descendants, wore heavy armour which must have restricted its movement, but which also made it virtually impregnable. Later in the Mesozoic other forms evolved, some of which took up a marine way of life. In the water the turtles could move faster, which meant that they were not so dependent on their armour. Their shells became lighter and they developed long, paddle-like feet. *Archelon*, a late Cretaceous turtle, was four metres (13 feet) long, but its shell was reduced to a few struts and plates of bone. Today, warm seas and remote islands still provide a natural habitat for some of the gigantic turtles and tortoises.

Right: Protosuchus, a direct ancestor of the crocodiles. Its strong legs show that it was far less sluggish than its modern relatives.

Below: Rutiodon, a phytosaur from the Upper Triassic. The nostrils, placed high up between its eyes, enabled it to float just under the surface of the water, waiting for its unwary prey.

THE RULING REPTILES

The chart shows the evolution and relationships of the archosaurs. They were probably all descended from a thecodont, like *Euparkeria*. The two main groups are the Saurischia and the Ornithischia.

Saurischia
The saurischian order comprises the sauropods – quadrupedal herbivores, such as *Apatosaurus* – and the theropods. The theropods were carnivorous bipeds and there were two groups of them – the smaller coelurosaurs, such as *Ornithomimus*, and the larger carnosaurs, typified by *Tyrannosaurus*.

Ornithischia
The ornithischians were all herbivores. One branch, the ornithopods, included bipedal forms, such as *Iguanodon*, and the later hadrosaurs, such as *Corythosaurus*. They also gave rise to the quadrupedal ceratopians, such as *Triceratops*. The other quadrupeds were the stegosaurs and their successors, the ankylosaurs.

The remaining archosaurs are the crocodiles and the extinct pterosaurs. The most successful descendants of the group are the birds, although their direct ancestry is unknown.

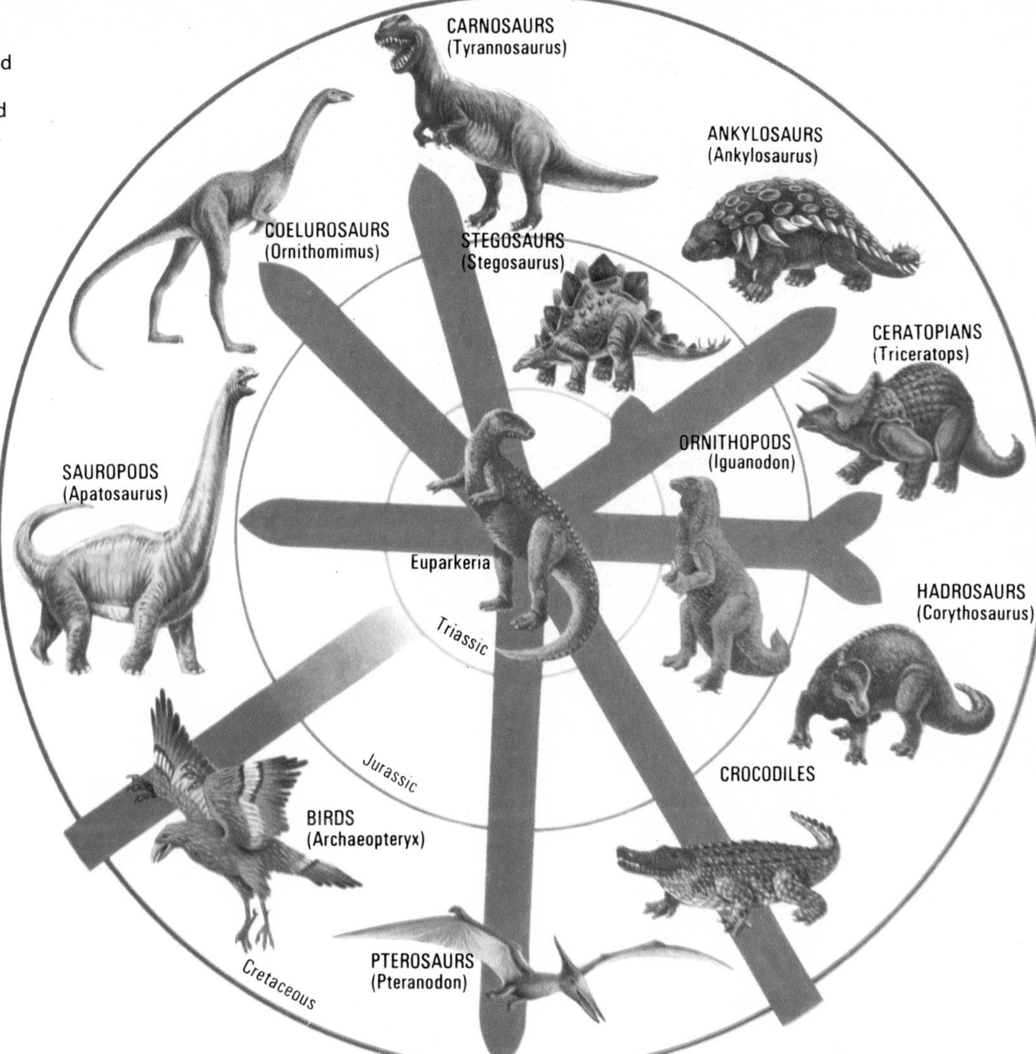

The ornithischian and saurischian dinosaurs are basically distinguished by the shape of the pelvic girdle. In the ornithischian pelvis (top), the pubis is parallel to the ischium. In the saurischian (bottom), it points downwards and forwards.

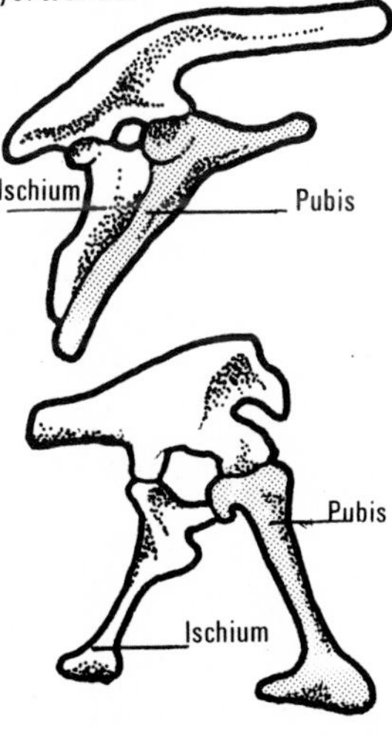

The 'Ruling Reptiles'

The only living members of the once dominant group of 'ruling reptiles', or archosaurs, are the crocodiles and alligators. With their grotesque appearance and aggressive habits, these animals have changed little since they evolved in the late Triassic. Lurking by the banks of lakes and rivers, they prey on any animals which may come to the water to drink. They escape detection by lying motionless and almost wholly submerged, with just their eyes and nostrils rising above the surface. Only the unsuspecting victims come to realize their presence as they feel the snap of the jaws and the huge teeth sinking into their flesh.

Although crocodiles rarely stray far from the shallows, the side-to-side lash of their powerful tails makes them strong swimmers. On land they are capable of a surprising turn of speed when frightened, but in general they can do no more than lumber about in an ungainly fashion. Their earlier relatives, however, were more agile. The long legs and ankles of *Protosuchus* suggest that it was capable of running quite fast.

Protosuchus was only about one metre (three feet) long. Later crocodiles were often much bigger. The late Cretaceous *Deinosuchus* was about 15 metres (50 feet) long; its head alone measured two metres (six feet). *Deinosuchus* must have been a formidable adversary even for a dinosaur.

The crocodiles were successful in the water as well as on land. A group of marine species including *Teleosaurus* and *Steneosaurus* invaded the seas in Jurassic times. They lost their bony armour, and instead developed paddles and a tail, like the ichthyosaurs.

Alongside the crocodiles, other groups of archosaurs evolved. Some, such as the

WHY SO BIG?

The most remarkable characteristic of the dinosaurs was their sheer size. Though there were some small dinosaurs – little *Compsognathus* was no bigger than a chicken – there is no doubt that most of these animals were large, if not enormous. The dinosaurs had no hair or feathers to regulate their body temperature, and their huge size was a protection against everyday temperature variations. As an animal grows larger, its surface area, through which heat is lost and absorbed, becomes proportionately much smaller. Fluctuations in body temperature therefore occur more slowly. Small variations – between day and night temperatures, for instance – would have had little effect on the hulking beasts. Only long spells of unusually hot or cold weather would affect such large animals. During the Mesozoic Era, the earth's climates remained almost unvaryingly mild and the dinosaurs basked in warm security. But this confidence was ill-founded, for it was probably their inability to cope with long-term temperature changes which proved to be their fatal weakness, when the world became cooler at the end of the Cretaceous Period.

A number of life-sized reconstructions of dinosaurs were made and placed in the grounds of the Crystal Palace in 1854. They were made by a sculptor called Waterhouse Hawkins, under the guidance of Richard Owen. These two men, together with 20 guests, had dinner one day within the half-completed body of Iguanodon!

phytosaurs *Rutiodon* and *Phytosaurus*, were as large and aggressive as the big crocodiles. Others, such as *Stagonolepis* – an aetosaur – were herbivorous, but still big and well-armoured. These creatures were all quadrupedal like the crocodiles, but *Euparkeria*, one of the earliest archosaurs, was able to run on its hind legs, using its tail to counterbalance the weight of its body. The phytosaurs became extinct at the end of the Triassic.

Terrible Lizards

Nobody knows for sure what the first dinosaurs were like, but they were certainly not as spectacular as their descendants. The earliest known dinosaurs lived in the late Triassic Period. At that time, today's continents were still united in a single land mass, and the remains of these early dinosaurs are widely distributed throughout the world.

The name 'dinosaur' means 'terrible lizard', and is popularly applied to the great land reptiles which make up the two extinct orders, Saurischia and Ornithischia (see chart on page 105). The animals of these two orders differ in the structure of their jaws and pelvic girdles. In fact, they are not closely related; they may even have evolved from separate stocks. Scientists used to believe that all dinosaurs were descended from a common bipedal ancestor similar to *Euparkeria*, and that quadrupedal members of both orders evolved much later. But many scientists now think that the earliest saurischian dinosaurs were quadrupedal.

Whatever their ancestry, the two orders contain a vast and fascinating selection

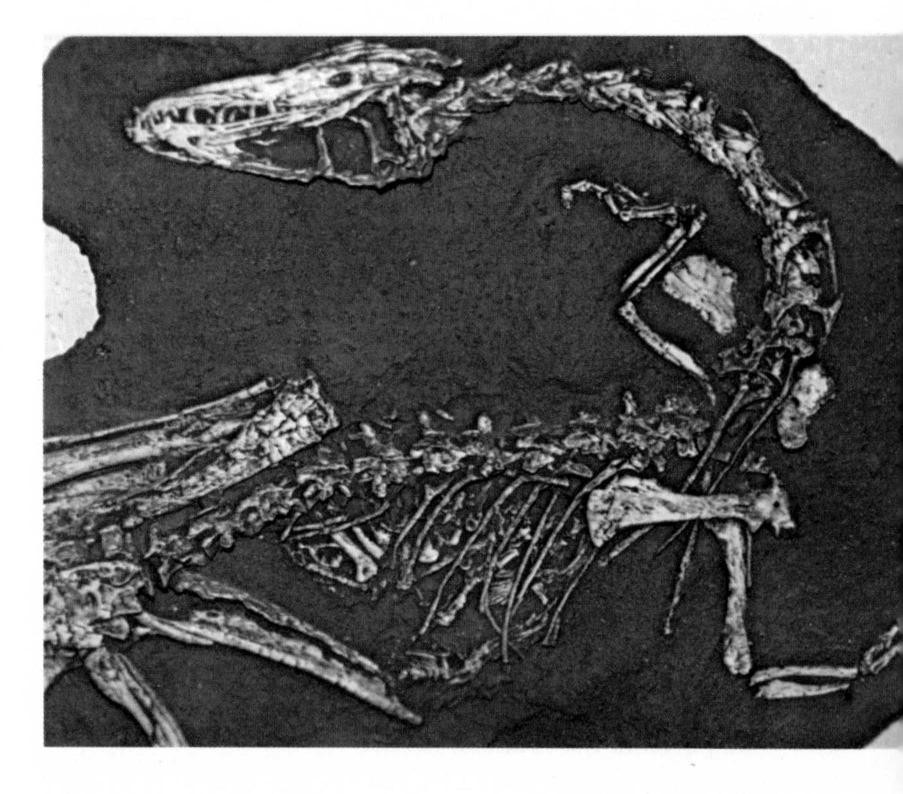
Below: Coelophysis, a Triassic theropod from North America. Scientists are puzzled by the presence of small Coelophysis bones within this specimen, since it is unlikely that the animal gave birth to live young or ate its own offspring.

RICHARD OWEN

In the early days of palaeontology, it was the anatomists who were best equipped to realize how unlike the remains of any living creature were the petrified objects which were then being recognized as fossil bones. Richard Owen (1804–1892), an anatomist and zoologist, set about a comprehensive study of the huge reptilian bones, which had recently been excavated in England and other parts of Europe. By 1841 he had become convinced that these were so totally different from the bones of any living reptile that they must have belonged to a group which was now extinct. He called this group the Dinosauria, a name which has passed into everyday use to describe the large reptiles that are now classed as the Saurischia and Ornithischia.

Owen became a leading British scientist. After holding a post at the Museum of the Royal College of Surgeons, he became the first director of the Natural History Museum in London. It is interesting that Owen did not fully accept Darwin's view of evolution. He believed that God had created each major group of animals, and that evolution had taken place only *within* each of these groups.

of animals. The various groups into which the orders are subdivided can be seen in the chart on page 105.

The First Dinosaurs

In any community of animals there are many more herbivores than carnivores. Thus, among the dinosaurs, carnivorous species – though fierce – were relatively few. All the ornithischians were herbivorous, and so were most of the saurischians. Only the theropods, a group comprising the coelurosaurs and the carnosaurs, were carnivorous.

One of the earliest known dinosaurs was a coelurosaur called *Coelophysis*. A slender creature, some two and a half metres (eight feet) in length, it ran on its hind legs and used its fore-limbs for grasping its prey. Another saurischian was *Plateosaurus*. This bulky herbivore was about six metres (20 feet) long. Plodding about on all fours, it would have had little chance of escaping from the fleet-footed carnivores. *Heterodontosaurus* was not such easy prey. Barely bigger than a goose, this little ornithischian biped could probably have outrun larger creatures like *Coelophysis* with ease.

The considerable differences in form and habits displayed by these few species show that, even in the Triassic, the dino-

Despite their bulk, the early saurischian herbivores were easy prey for the Triassic carnivores. The slow-moving Plateosaurus could not easily escape the tenacious teeth of the thecodont, Ornithosuchus. ***As the herbivores grew larger, however,*** *the thecodont carnivores were replaced by bigger and fiercer theropod dinosaurs.*

saurs were evolving in many different directions. The warm climates of the Jurassic produced an even more hospitable environment. Vegetation was abundant and a rich array of dinosaurs emerged to occupy the wide range of new ecological niches.

Although shallow seas had invaded many parts of the world, links still existed between the continents. Fossils from the Rocky Mountains in North America and from sites in East Africa show that a similar dinosaur fauna inhabited both the northern and the southern hemisphere.

Giants of the Jurassic

One dinosaur which has been discovered in both regions is the giant *Brachiosaurus*, which belonged to a group known as the sauropods. This enormous beast was 24 metres (80 feet) long and weighed as much as 51 tonnes (50 tons). With its neck held erect, it must have been able to browse happily from the topmost branches of 12-metre (40-foot) trees. Though *Brachiosaurus* holds the record as the largest land animal which has ever lived, it only just outstrips in volume some of the other huge sauropods, such as *Apatosaurus* (*Brontosaurus*) and *Barosaurus*. The bulky bodies of the sauropods tapered into a long neck at one end and a long tail at the other, making them the longest as well as the heaviest reptiles. *Diplodocus* of the late Jurassic measured 27 metres (90 feet) in length.

A scene from the late Jurassic. In the foreground, the small coelurosaur, Ornitholestes (left) is chased by Allosaurus (right). The huge carnosaur would have had little difficulty in frightening away any rival that strayed into its path. In the background, the herbivores munch happily, heedless of the conflict. Camptosaurus (far left) rears up on its hind legs to reach the juicy leaves above its head. With little effort, Brachiosaurus (centre) browses on the very topmost branches while Stegosaurus (right) crops the ground cover.

Like modern herbivores, the sauropods must have spent almost all of their time feeding. Their heads were relatively so small that it seems almost inconceivable that they could ever have taken in enough food to assuage the hunger of their colossal bodies. But the sauropods obviously needed little energy. Protected by their sheer enormity from all but the most intrepid carnivores, they can seldom have been compelled to take violent exercise.

Predators, Large and Small

Allosaurus was one of the few animals which might have dared to attack *Apatosaurus*. This vast, ferocious reptile belonged to a group known as the carnosaurs. It was more than nine metres (30 feet) long, and covered the ground in huge strides. Its clawed fore-limbs were used for grasping prey. Its head was enormous and its jaws and teeth so strong that it must have been able to tear up the smaller dinosaurs in minutes.

LAND OR LAKE DWELLERS?

Scientists believed until recently that the great sauropod dinosaurs lived in lakes, because it seemed impossible that they could have supported their own weight on land. Their long necks were thought to have evolved to allow them to keep their bodies submerged in deep water, safe from the attacks of carnivores, while still keeping their heads above the surface. Their nostrils are usually very large and placed rather far back, and this was interpreted as a device which would allow them to breathe with only the top of the head exposed.

It now seems, on the contrary, that the sauropods were normal, terrestrial animals. For example, their limbs do not end in expanded feet like those of living animals (such as hippos) which live in soft, muddy ground. Instead, they had round pads, like those of elephants, which quickly become mired and trapped in mud. The limbs of sauropods were also like those of elephants – strong, straight pillars capable of carrying heavy body weights. Their body shape, too, is deep, unlike the barrel-shaped hippopotamus.

It is in any case difficult to see the value of a long neck to an aquatic sauropod. If they had ventured into such deep water that they needed long necks to reach the surface, the pressure of the water above their bodies would probably have made it very difficult for them to expand their chests and draw in air. Anyway, there would have been too little vegetation at such depths to have provided the food requirements of a sauropod. A long neck would also have been of little use on land, if it had been held horizontally, as depicted in many older reconstructions. It seems far more likely that the advantage of an elongated neck was that it enabled the animals to feed on the topmost branches of trees, as modern giraffes do – in which case, the neck would have been held upright.

Contrary to popular belief, the huge sauropod dinosaurs were almost certainly land dwellers. Though a creature such as Barosaurus would rarely have had cause to move fast, it is obvious from its bones that it held its neck upright and could gallop away from danger like a reptilian giraffe.

During the Cretaceous Period, some even more terrifying carnosaurs emerged. *Tyrannosaurus*, *Tarbosaurus*, and *Gorgosaurus* were the largest flesh-eaters ever to live on land. *Tyrannosaurus* was 16 metres (52 feet) long and stood six metres (20 feet) high on its hind legs. Its fore-limbs had dwindled almost to vestigial structures; they bore only two claws and were quite useless for support or feeding. But with great claws on its feet and a grim mouthful of dagger-sharp teeth, *Tyrannosaurus* must have been a fearsome predator.

The coelurosaurs were smaller, if no less offensive creatures. Like *Coelophysis* of the Triassic, they could speed about on their long, nimble legs and grasp their prey with their flexible, clawed fingers. Creatures such as *Ornitholestes* probably also scavenged on the kills abandoned by larger carnivores. The most elegant of all the Cretaceous predators was *Ornithomimus*, a creature more like an ostrich than a dinosaur. It had legs like stilts and a tiny beaked head, mounted on top of a long, slender neck. Better equipped for stealth than combat, it probably survived on a diet of insects, lizards and eggs stolen from the nests of other dinosaurs.

Armoured Pacifists

The giant sauropods and the great carnivores belonged to the saurischian order of dinosaurs. But, although this group takes the palm for size, it certainly cannot match the ornithischians for variety and innovation. The latter order embraces some of the most bizarre and highly specialized animals which have ever lived.

One of the most famous of the armoured dinosaurs, *Stegosaurus*, lived during the Jurassic Period and fed on the lush vegetation which then carpeted the ground. *Stegosaurus* was about seven metres (23 feet) long, and its hind limbs were almost twice the size of its fore-limbs. Its ancestors were probably bipedal creatures that had gradually dropped down on to all fours to support their weight.

Unable to run very fast, *Stegosaurus* adapted itself for another form of defence against its foes. Its tail was barbed with long, bony spikes. One blow from it would have incapacitated all but the most determined challengers. At close quarters, *Stegosaurus* was as well fortified as a mediaeval castle. Its neck and back were protected by an upright line of triangular bony plates. Perhaps its flanks were the one weak spot in its defences: the leathery skin was encrusted with bony lumps, but even so this part of the body would have proved vulnerable in a lengthy tussle. Despite its armour, *Stegosaurus* was no match for the carnivores that preyed on it, and by the end of the Jurassic this placid plant-eater had disappeared forever.

During the Cretaceous Period the position left vacant by *Stegosaurus* was taken up by a slightly smaller but even more daunting dinosaur called *Ankylosaurus*. Its body, supported on short, sturdy legs,

ROCKY MOUNTAIN FOSSILS

The Rocky Mountains in North America arose in the Jurassic and Cretaceous Periods. Sediments from the erosion of these mountains were soon being deposited in flat, lowland areas to the east. These sediments later hardened into the rocks of the Morrison Formation, from which we get our best view of dinosaur life in the Jurassic. Many skeletons and bones were buried in the flood-plains and preserved as fossils. They are found over wide areas of Utah, Colorado, Wyoming and Montana.

NOT SO BRAINY?

Stegosaurus is often said to have had two brains – one in its head and a much larger one in its tail. In fact, the tiny one in its head was its only one. The 'brain' at the back was simply an enlargement of the nerve chord at the base of the spine – no doubt useful for controlling the movement of its hind limbs and tail.

A warden from the Dinosaur National Monument stands beside the bones of an enormous dinosaur embedded in the rock face. The national park covers an area of Colorado and Utah, where the rocks of the Morrison Formation yield their spectacular array of Jurassic fossils.

THE RIVALS

Othniel Marsh (1831–1899) was the first professor of palaeontology at Yale and curator of the Peabody Museum which he set up with the backing of his rich financier uncle, George Peabody.

In western America, where vast tracts of Jurassic and Cretaceous sediments lie exposed, fossils are so abundant that a shepherd in Wyoming once made himself a cabin from the enormous fragments of bone. But even this abundance of land and fossils could not contain peacefully the rival ambitions of two American palaeontologists, Edward Drinker Cope (1840–1897) and Othniel Marsh (1831–1899). Both were dedicated zoologists and wealthy men, passionately interested in the discovery, assembly and description of the dinosaurs whose scattered bones were waiting to be dug from the rocks.

The two men became bitter enemies. Each eagerly attempted to use his wealth to corner the market in dinosaurs, and rival teams were sent out west, pledged to keep secret all their specimens and collecting sites.

Spurred on by this rivalry, the men managed to identify more than 130 species (although some of these were simply rival accounts of the same animal). Among the Jurassic dinosaurs they found were *Allosaurus, Brontosaurus, Stegosaurus, Diplodocus* and *Camptosaurus.* From the Cretaceous they found *Hadrosaurus, Ornithomimus, Triceratops* and *Nodosaurus.*

Edward Drinker Cope (1840–1897) was an indefatigable researcher all his life. Part of his collection of fossils can be seen in the American Museum of Natural History.

was broad and fairly flat. Over its back it wore a leathery cloak, embroidered with bony plates and trimmed with spikes. The end of the tail bore a heavy cudgel which could inflict a limb-cracking blow. If this defence failed, the animal could drop on to its stomach. In this position its armour formed a stronghold capable of withstanding savage assault.

Early Ornithopods

The ornithopods were a group of ornithischians which appeared during the Triassic. *Heterodontosaurus* and *Lycorhinus* were two of the earliest. The ornithopods were the only truly bipedal herbivores. *Camptosaurus*, like the other early types, probably walked on all fours, but to run or get food it reared up on its hind legs. In this position it could reach the higher branches of trees, like the modern gerenuk. *Camptosaurus* was well adapted to a diet of leaves and other vegetation. It had a toothless beak for cropping foliage and a battery of flat, slicing cheek teeth for cutting it into pieces. Although relatively small – it was between two and six metres (six and a half and 20 feet) long – *Camptosaurus* was nonetheless an evolutionary success. Most of the Cretaceous ornithischians were descended from it.

Gregarious Herbivores

Iguanodon, a bigger and bulkier ornithopod of the early Cretaceous Period, was even better equipped for a herbivorous life. It had several rows of teeth side by side; as the older teeth were worn down, new ones grew up on the inside to replace them.

Completely bipedal, *Iguanodon* was more than nine metres (30 feet) long, with a heavy tail and large, muscular hind limbs. Its fore-limbs were fairly small, but its thumbs were modified into sharp, bony spikes, which may have had a defensive function in hand-to-hand fighting. On the whole, no doubt, *Iguanodon* relied on

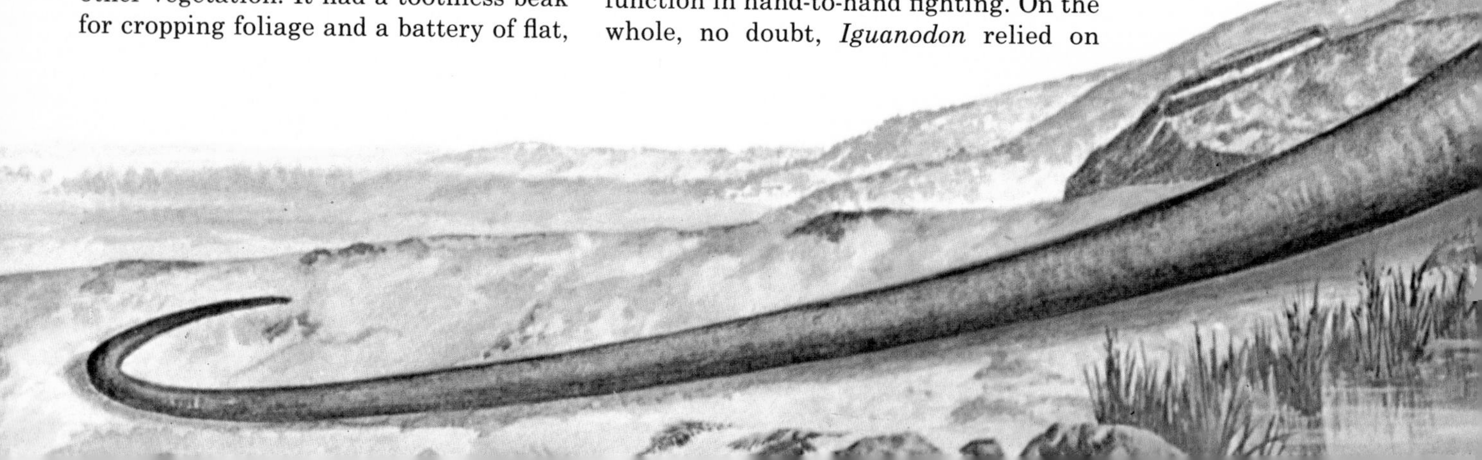

Diplodocus has the record for being the longest land animal ever to live. It was 27 metres (90 feet) long. With a tiny head and surprisingly puny teeth, it must have spent all its time trying to gather enough food to fuel its giant body.

speed to escape from enemies, as well as on the protection afforded by the herd. It has always been assumed that herbivorous dinosaurs, like modern herbivores, generally lived and foraged in herds; but in the case of *Iguanodon* we can be quite certain of this. A herd of 20 of the dinosaurs, presumably buried by a landslip, was discovered during excavations in a Belgian coal mine.

The Mysterious Duck-Bills

The hadrosaurs were a group of very successful ornithopods which lived in the late Cretaceous Period. Specimens have been found that are almost as well preserved as Egyptian mummies, with skin as well as bones intact, so that we have a very vivid idea of their appearance. Although the basic structure of the hadrosaurs fits neatly into the ornithopod pattern, many of their adaptations are so bizarre that scientists can find no convincing explanation for them.

The animals were mostly large, some nine to twelve metres (30 to 40 feet) long, with big hind limbs and broad feet. One strange adaptation has earned them the name of 'duck-bills'. The skull and the lower jaw formed a broad, horny, toothless beak. In *Anatosaurus* and *Lambeosaurus* these beaks were especially pronounced. Inside the jaws was an array of up to a thousand tightly-packed teeth. These formed two huge grindstones between which the food was pulverized.

The most enigmatic feature of the hadrosaurs is the variety of 'crests', of all shapes and sizes, which bedecked their heads. These crests were enlargements of the nasal region of the skull. In *Parasaurolophus* it took the form of a hollow bar sweeping back over the head. In *Corythosaurus* it was helmet-shaped and in *Saurolophus* it was horn-shaped. *Anatosaurus* was crestless but *Lambeosaurus* had both a helmet and a horn-like structure projecting backwards from the head. The specific function of these crests is unknown. They may have enhanced the animals' sense of smell, or made their bellows more resonant.

MANTELL AND IGUANODON

The skeleton of Iguanodon, a lower Cretaceous ornithopod dinosaur some nine metres (30 feet) or more long.

Gideon Mantell (1790–1852) was born in Sussex, where fossils (especially invertebrate fossils) are not uncommon. Even as a boy he was interested in these. Later he became a doctor, and in 1822, while he was visiting a patient, his wife – taking a walk while she waited for him – noticed some large fossil teeth. Huge bones were later found in similar rocks in the same area. After many enquiries and comparisons, Mantell eventually realized that these teeth looked very like those of the living iguana. He therefore named the fossil *Iguanodon* – 'iguana tooth'. It was the second dinosaur species ever described.

It was not until 1834 that a partial skeleton of *Iguanodon* was found. This made it possible to attempt one of the first-ever dinosaur reconstructions. Inevitably, it was wrong in some respects – Mantell believed the animal to have been quadrupedal, not bipedal, and the characteristic thumb-spike was thought to be a rhino-like horn. But at least the general public, as well as the scientific world, was coming a little closer to understanding just how unlike any living animal these gigantic reptiles were.

Below: A scene from the late Cretaceous of North America. In the foreground, Tyrannosaurus, the largest and most ferocious land carnivore ever known, seizes the helmeted hadrosaur Corythosaurus by the throat. In the background, another Corythosaurus (left) prepares to flee before meeting the same fate, while Ankylosaurus (centre), safe under its armour, can continue to browse. In the foreground Triceratops (far right) gallops away, while the audacious coelurosaur Ornithomimus sneaks up to raid another dinosaur's nest.

THE CRETACEOUS SCENE

The dinosaurs of the Cretaceous lived in a world whose geography was changing rapidly. During this period nearly all the present-day continents appeared as separate land masses. The northern and southern continents became almost completely isolated from one another, and in the late Cretaceous shallow seas crossed North America and Eurasia from north to south, dividing the northern supercontinent into two areas of land quite different from those which exist today. One was made up of Asia and western North America, which were connected via Alaska. It was in this continent that most of the new dinosaur types of the Cretaceous appeared. The other continent was made up of eastern North America and Europe, which were connected via Greenland. In the mild environment of the Cretaceous, even Alaska and Greenland had warm climates and provided possible migration routes for land vertebrates.

The Horned Dinosaurs

During the late Cretaceous the dinosaurs achieved their final evolutionary flourish with the brief appearance of the ceratopians, or horned dinosaurs. While the ornithopods seem to have filled the niches occupied by antelopes and deer in the modern world, the ceratopians were more like rhinoceroses – well-armed and immensely powerful.

The armour of the first small ceratopians, such as *Protoceratops*, consisted of an enlargement of the skull in the form of a hooked beak, and a frill of bone like a turned-up coat collar around the head. This served to protect the neck and shoulders. In later ceratopians the frills became more elaborate, and were spiked with cruelly sharp horns. *Monoclonius* had a single horn on its nose, while *Triceratops* and *Chasmosaurus* each had one on the nose and another two on the forehead. In *Styracosaurus* the whole frill was fringed with spikes.

The largest ceratopians weighed as much as eight and a half tonnes (eight tons) and were more than seven and a half metres (25 feet) long. Few animals could have survived a charge from one of these brutes and even the largest carnivores must have shrunk from close combat.

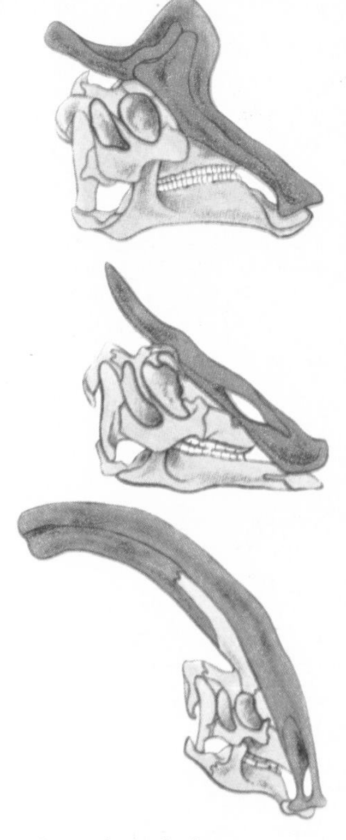

Above: Some hadrosaur crests (top to bottom: Lambeosaurus, Saurolophus and Parasaurolophus). The function of these bizarre structures remains a mystery.

The Boneheads

A head-on collision with either *Pachycephalosaurus* or *Stegoceras* must have been a terrifying prospect. Little is known of these animals except that above their brains they carried a solid mass of bone over 25 centimetres (10 inches) thick. In *Pachycephalosaurus* this dense area was surrounded by a number of bony studs. The function of this cumbersome appendage is uncertain. It may have been used as a battering ram when *Pachycephalosaurus* joined battle with an enemy, or when it fought over a female with others of its own species. The block of bone would then have ensured that the animals did not crack each other's skulls during the fighting.

Above: Like a permanent crash helmet, Pachycephalosaurus carried a dense mass of bone over its head. This protection would have ensured that the animals did not bash each others' brains out when engaging in courtship battles with other males.

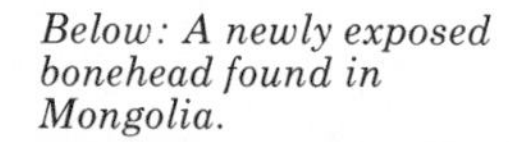

Below: A newly exposed bonehead found in Mongolia.

The Death of the Dinosaurs

At the end of the Cretaceous, the dinosaurs seemed set to continue their dominance for another 150 million years. They had adapted to the new flowering plants and appeared to be flourishing. Yet only a few million years later, all were dead – herbivore as well as carnivore, ornithischian as well as saurischian, the merely large along with the enormous. What could have caused so comprehensive a catastrophe?

Many theories have been advanced to explain the extinction of the dinosaurs. Some are hard to believe; others impossible to prove. It has been suggested that a dramatic increase in solar radiation was responsible. Alternatively, a reversal of the earth's magnetic field might have caused a temporary breach in the ozone shield which protects the planet from harmful cosmic rays. Others again believe that alkaloids developed in plants and poisoned the dinosaurs, or that the great beasts were wiped out by the spread of bacterial disease.

For a long time scientists thought that the dinosaurs might have been the unfortunate victims of egg-eating mammals. But the innocuous little mammals had been in existence ever since the dinosaurs first appeared. Indeed, they seem to have had no chance of success until after the dinosaurs became extinct. Again, competition from the mammals would, at best, only explain terrestrial extinctions. Yet it was not only the dinosaurs that disappeared. At the end of the Cretaceous, ichthyosaurs, plesiosaurs, ammonites, and almost all creatures of any size vanished from the seas. Palaeontologists have found it difficult to formulate any theory which would explain both terrestrial and marine extinctions.

It has been suggested that the initial cause of the marine extinctions was a major decline in the numbers of planktonic plants during the late Cretaceous Period. These plants form the basis of the marine food chain, and their disappearance would almost certainly have set off a reaction throughout the entire system. It is, however, unlikely that this would have affected the land fauna in any serious way. Similarly, an increase or decrease in oceanic salinity is unlikely to have been responsible, though an increase in the salinity of the oceans has been suggested as the cause of the Permo-Triassic extinctions.

The most plausible explanation proposed so far is that the animals could not survive the drastic climatic and environmental changes which occurred as a result of continental drift. As we have seen, the one weak spot in the dinosaurs' defences was their dependence on a consistently mild climate. Since they lacked any kind of insulation, they could not protect themselves against extremes of temperature. At the end of the Cretaceous Period, climates must have become harsher. Unable to adapt to prolonged cold seasons and to greater annual variations in temperature, the dinosaurs declined and died out.

This abrupt change of climate was brought about by a lowering of sea levels. Oceans widen or shrink as a result of continental drift. As the continents move apart, new sea floor is produced in the great oceanic ridges. The ridges grow hot and expand, thereby reducing the capacity of the ocean basins. The oceans overflow on to the lowest-lying areas of land, forming shallow seas; and it is in these shallow seas that marine animals mostly thrive. The presence of seas also has a profound effect on climate, giving rise to milder weather with less marked seasonal extremes.

For most of the Mesozoic the continents were continually on the move, and shallow seas flooded the land. But at the end of the Cretaceous, continental drift suddenly slowed down, and the seas withdrew from the continents back into the ocean basins. The resulting change of climate could easily have wiped out the dinosaurs and decreased living space in the sea might well have led to the extinction of many marine animals.

NEST EGGS

A lucky find by palaeontologists has provided us with a detailed record of ceratopian life history. Nests of eggs, together with specimens of *Protoceratops* at various stages of development, have been excavated in Mongolia. The nests – simple holes scooped in the sand – contained rings of thick-shelled eggs. Like modern turtles, the females must have laid their eggs, covered them with sand, and left them to hatch in their own good time.

The Mystery Remains

Despite the plausibility of the continental drift theory, it does not answer all the questions. It is surely possible that the dinosaurs could have moved to the few inland areas where equable climates persisted, and that some at least of the marine reptiles could have survived in the shrinking seas. Such sweeping environmental changes would also normally be expected to affect all species; yet the extinctions were selective. Certain animals survived and certain others did not. Differences in physiology are obviously responsible for the survival of some groups – the mammals, for instance, which had hair, and the birds, which had feathers. But why did the crocodiles, alone among the archosaurs, survive? Did they possess some fundamental but so far unrecognized adaptation? Or is it just that, like the surviving chondrostean fishes, they have found a niche to which they are extremely well adapted?

A ring-shaped nest of eggs belonging to the ceratopian dinosaur Protoceratops.

The Conquest of the Air

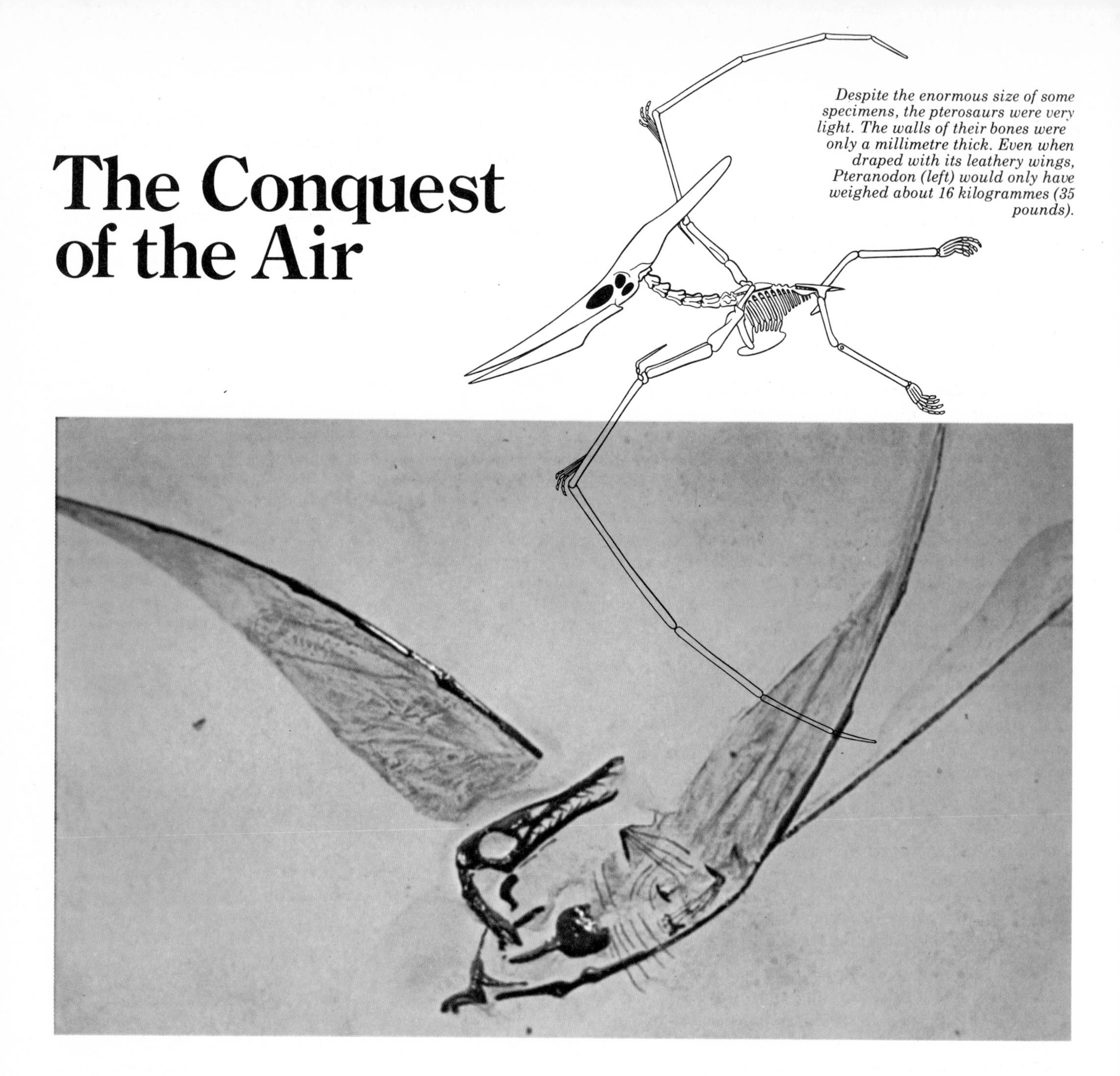

Despite the enormous size of some specimens, the pterosaurs were very light. The walls of their bones were only a millimetre thick. Even when draped with its leathery wings, Pteranodon (left) would only have weighed about 16 kilogrammes (35 pounds).

Rhamphorhynchus was a fairly small Jurassic pterosaur. The fossil shows the elongated tail with its rudder-like end and the long, thin jaws. The forward-pointing teeth may have helped the animal to spear its prey of fish as it skimmed over the water.

Ground-living animals are in constant danger of discovery and attack by hostile carnivores. Their eggs, too, are at the mercy of predators. The evolution of the power of flight can therefore be seen as a far-reaching self-defence policy on the part of the terrestrial archosaurs. Once they had 'learnt' to fly, they could lay their eggs with confidence in the treetops, on the ledges of sheer cliffs, or on off-shore islands. They could also exploit new sources of food. Flying insects had become much more common in late Palaeozoic times; and fish in the surface layers of the sea were easy prey for any animal that could swoop down and snatch them.

There are two ways in which the flying habit might have evolved. Some of the ground-living reptiles might have taken to the air as they ran to escape a predator, gaining lift by flapping their fore-limbs; or certain tree-living species, which could glide to the ground with the help of a 'bat's-wing' type membrane, might have acquired the ability to propel themselves forward while remaining airborne. The second theory seems the more plausible. Even today, some small mammals and reptiles glide from tree to tree, using a membrane attached to the fore-limbs to extend their body surface and thus supply the necessary lift. Such behaviour might well have led to the development of a structure which could also be used for flight. Ground-living animals, on the other hand, would not have been able to gain height fast enough to shake off a predator until effective wings and flight muscles had evolved – and such equipment is a very specialized development.

Gliding is not a physically demanding activity; but it does require an ascending current of air, such as is found near cliffs or over areas heated directly by the sun. Flying animals obviously have greater freedom of movement, but they must use up energy continuously in order to stay in the air. The ancestors of the birds were thus forced to develop an effective system of insulation so that they could retain their body heat and use it as a source of energy. This is the function of the feathers in modern birds. Like the mammals, another physically active group, birds are warm-blooded animals.

Fish-Eating Dragons

The marine sediments of the Jurassic and Cretaceous have produced a variety of skeletons belonging to an extinct group of gliding reptiles, the pterosaurs. Circling majestically above the surface of the sea, swooping down occasionally to impale fish, and returning to perch, bat-like, on the side of a cliff, these animals were cumbersome creatures compared with the birds of today. But they possessed many of the same bodily adaptations for flight.

The pterosaurs were related to the dinosaurs. At one extreme the group included tiny specimens the size of a sparrow; at the other, giants like *Pteranodon*, whose wingspan was nearly seven metres (23 feet), and an immense Texan specimen, whose wings spread 15·5 metres (51 feet) from tip to tip. But hollow bones, with walls only a millimetre thick, made the pterosaurs very light in relation to their size. The chest bone could not support muscles capable of flapping the enormous wings; these must have been used mainly for manoeuvring. One species, *Rhamphorhynchus*, also had a vertical 'rudder' at the end of its long tail. Another typical feature of the pterosaur skeleton was the long skull, armed with sharp forward-pointing teeth for fish-catching (as in *Rhamphorhynchus*) or with a horny beak (as in *Pteranodon*).

Above: Dimorphodon, one of the most primitive pterosaurs, had a relatively large skull and a mouthful of hard, biting teeth.

Bat Wings

In some pterosaur fossils from the fine shale deposits of Solnhofen, in Bavaria, impressions of the wings are preserved. These seem to have been rather like bats' wings – leathery membranes which extended along the sides of the body to the hind legs. The front edge of the wing was attached to the arm, especially to the very elongated fourth finger. *Pteranodon*, with its light weight and large wings, would have been an excellent slow glider, well suited to the low wind speeds that must have prevailed in the mild climates of the Mesozoic. Models suggest that its airborne speed ranged from seven to 14 metres (32 to 46 feet) per second.

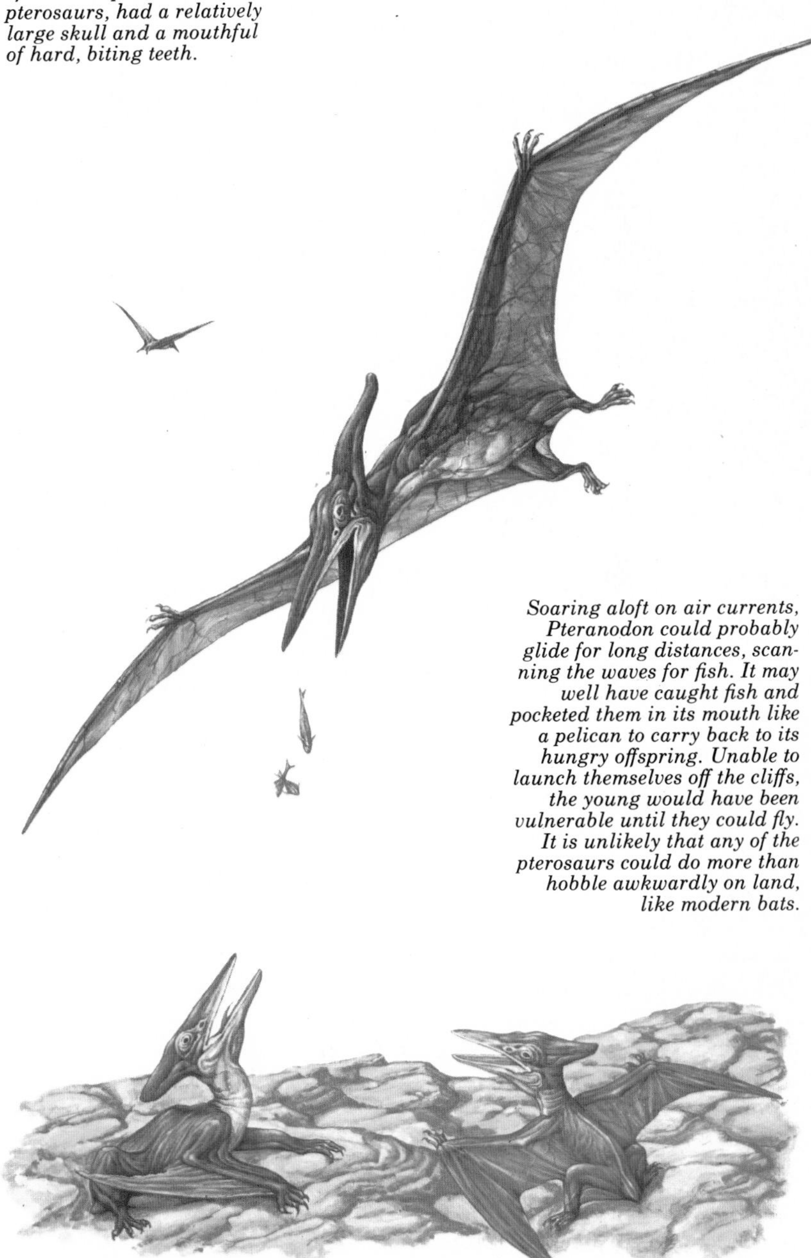

Soaring aloft on air currents, Pteranodon could probably glide for long distances, scanning the waves for fish. It may well have caught fish and pocketed them in its mouth like a pelican to carry back to its hungry offspring. Unable to launch themselves off the cliffs, the young would have been vulnerable until they could fly. It is unlikely that any of the pterosaurs could do more than hobble awkwardly on land, like modern bats.

An early stage in the evolution of birds, Archaeopteryx (above) still shows reptilian features. The tail was long and bony and the wings bore claws to help the bird climb. Its scaly head contained a beakful of teeth, but, as the fossil (opposite) shows, the rest of its body was clothed in feathers – the evidence that clearly marks Archaeopteryx as a bird.

On land, however, the pterosaurs must have been clumsy, almost helpless creatures. With their great wings folded across their backs, they must have supported themselves on their three projecting fore-fingers, using their hind limbs to push themselves forwards on their bellies. When not at sea, the pterosaurs probably lived on sea cliffs where they could hang suspended by their hind limbs. With no chance of building up speed for take-off like an aeroplane, the animals had to rely on the winds that commonly blow up cliff faces. They probably became airborne simply by dropping from the cliff edges.

Forced From the Sky

The pterosaurs seem to have possessed a body covering of hair for insulation. But they were not warm-blooded, and consequently not very energetic. They did not survive the competition of another group of archosaurs that had evolved wings about the same time. Perhaps the pterosaurs also found it difficult to cope with the strong winds which accompanied the cooler climates of the Tertiary. By the end of the Cretaceous all the flying reptiles were extinct. Their place in the sky was taken over by the birds, which had a far superior range of altitude, speed and distance.

Feathers and Flight

The Bavarian limestone quarries were destined to supply another vital missing link in the story of vertebrate evolution. In 1861, a fossil skeleton of the Upper Jurassic Period was discovered there. The fossil might easily have been accepted as belonging to a small dinosaur such as *Coelophysis* (see page 106). But distinctive traces in the surrounding rock made it clear that this creature had been covered with feathers. Closer study of the skeleton proved that it represented a very early stage in the evolution of birds. It was given the name of *Archaeopteryx*.

Feathers are a great improvement on the leathery flight membranes of pterosaurs and bats. Such membranes are difficult to repair and awkward to fold up when not in use, which makes their owners clumsy and ungainly when moving on land. In birds, however, the flight surface is a compound structure, formed of many separate elements – the feathers – which are difficult to damage but easy to repair or replace. These smaller units also enable birds to fold up their wings compactly when resting, so that they retain their freedom of movement in the trees or on the ground. Feathers, which almost certainly evolved from reptilian scales, also provide insulation, enabling birds to regulate their body temperature.

Below: The skeleton of Archaeopteryx (top) compared with that of a modern bird.

Claws For Climbing

Archaeopteryx has one very obvious feature which it shares with modern reptiles rather than with birds. This is its long tail, which bore a fringe of large feathers on either side. Again, the body of *Archaeopteryx* is longer than in modern birds, and its hand represents only an early stage in the evolution of the wing. The first two fingers still bore hooked claws, which the animal probably used for scrambling about in the trees. Unlike modern birds with beaks, its long jaws were filled with small, sharp teeth. Finally, *Archaeopteryx* had not yet evolved a sturdy, projecting chest bone to support its flight muscles; but it did have a long, wide tail. Both these facts suggest that, like the pterosaurs, it was more of a glider and less of a flyer than modern birds.

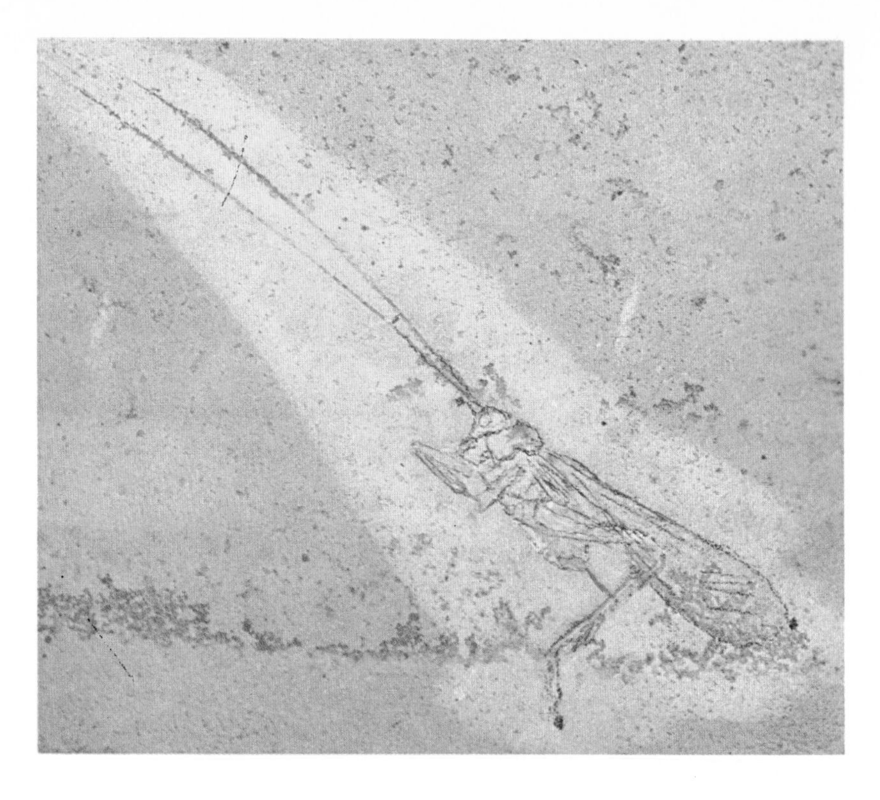

Above: The first insects were wingless but they must soon have evolved wings to help them escape predators and find food. This Eocene insect (Gryllacris) is a primitive member of the group that contains grasshoppers and locusts – the latter being among the best flyers of the insect world.

Below: Hesperornis, a Cretaceous sea bird, paddles through the water. Unlike the watching Ichthyornis (left), Hesperornis was incapable of flight.

Giant Water Birds

Fossil evidence from the chalk deposits of the late Cretaceous shows that, fifty million years after the time of Solnhofen *Archaeopteryx*, large numbers of fish-eating birds had made their homes in coastal regions all over the world. One of the most obviously primitive features of these Cretaceous birds is that, like *Archaeopteryx*, they still had teeth – useful equipment for catching and holding their slippery prey. Some species had already become so wholly marine that, like penguins, they had lost all powers of flight. *Hesperornis*, for example, retained only vestigial wings, and must have propelled itself through the water by powerful strokes of its hind limbs. Its smaller companion, *Ichthyornis*, was very like living seagulls; it must have dived into the water to seize its prey.

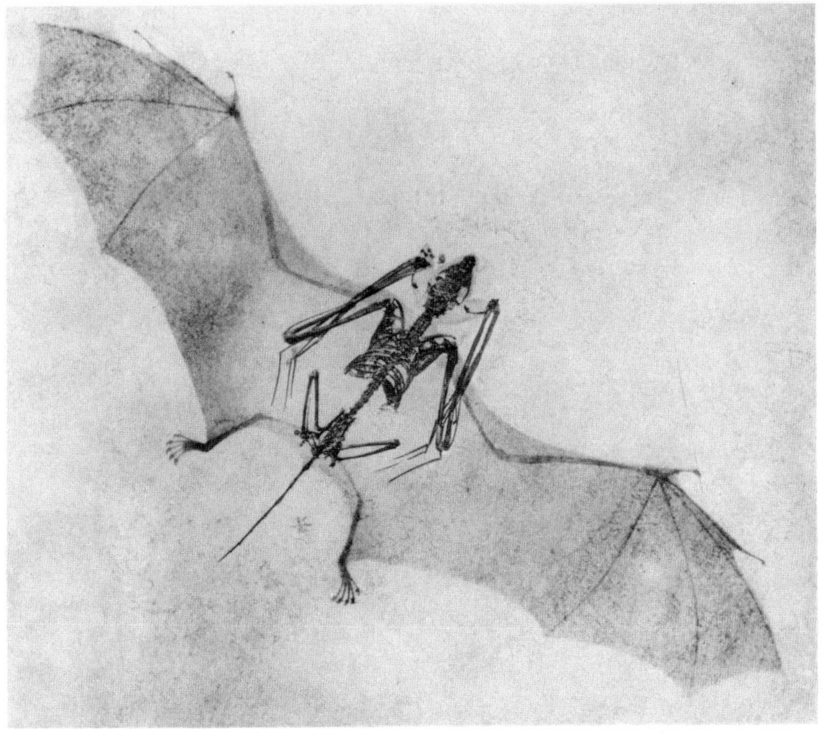

The commonest group of flightless birds, known as the ratites, seems to have evolved in Gondwanaland during the Cretaceous and to have colonized the whole of that supercontinent before it broke up. As a result, its living descendants are widespread – they include the rhea in South America, the cassowary and emu in Australia, the kiwi in New Zealand and the ostrich in Africa.

Various other giant ground birds gave up flying for running during the Tertiary. Two of the more important species were *Diatryma* and *Phororhachos,* neither of which was related to the ratites.

Above: The fossil of Icaronycteris, an Eocene bat. Superimposed on the fossil is a reconstruction of the outstretched wings.

Running Birds

Most birds rely on their powers of flight as a means of escaping from predators or of reaching secluded nesting sites. But flight is a very energetic method of locomotion, and some birds have found it more economical to abandon the habit and take to their legs. There are several groups of large, carnivorous flightless birds. These may originally have evolved in an area free from larger predators; but they eventually grew to such a size that they feared no enemies.

FLYING MAMMALS

The bats are the only mammals that have learnt to fly. Unlike birds, bats are helpless on the ground. Their clawed feet are adapted only for hanging upside down when asleep. The elongated bones of the fore-limbs form a frame, over which a skin is stretched to produce a structure like a child's kite. The bats display various adaptations for feeding. Some feed on insects, some on fruit, some on nectar – and some on blood from other animals.

All the bats fly out at night, and instead of sharp vision they have developed a hearing mechanism as acute as modern radar. In order to detect their prey, they emit a rapid series of short, very high-pitched squeaks through the nose. These are reflected back from any solid object and are picked up by the bats' enormously developed ears.

The earliest known bat is *Icaronycteris,* of which there is a beautifully preserved specimen from the Eocene of Wyoming.

Right: The powerful beaks and clawed feet of the giant flightless birds made them formidable adversaries. Diatryma (bottom), from the Eocene of North America and Europe, was more than two metres (six and a half feet) high. Phororhachos (top), with a head as large as that of a horse, lived in South America from the Oligocene to the Pliocene. Both birds seem to have become extinct when the larger carnivorous mammals evolved and spread into their homelands.

EPOCHS OF THE CENOZOIC

Because the Cenozoic Era covers the most recent part of the earth's history, there has been little time for rocks deposited in the Cenozoic to be destroyed by erosion or transformed by volcanic activity. This era has been a time of fairly rapid alteration in the climates and geography of the world, and a number of changes have therefore occurred in the faunas and floras of the different continents. For this reason the Cenozoic has been divided into seven distinct epochs. The most modern, called the Holocene or Recent, began only about 11,000 years ago, but the other epochs each lasted for between five and twenty million years.

The Cenozoic World

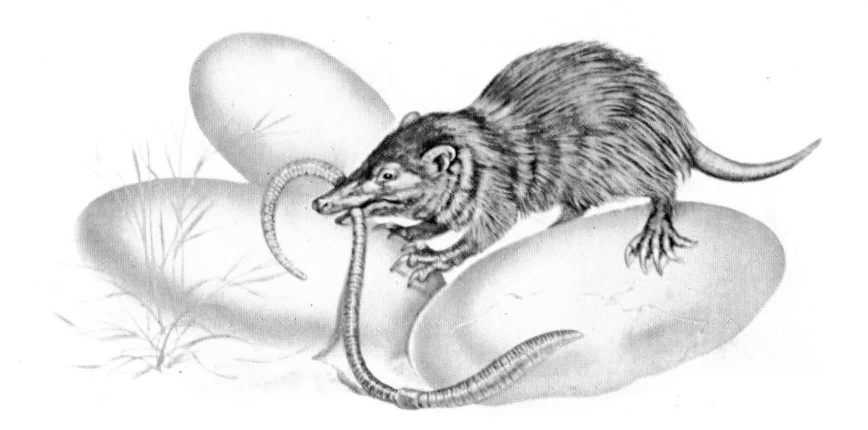

Right: Mesozoic mammals, such as Morganucodon, were hardly bigger than the eggs of the great dinosaurs. Only in the Cenozoic world did the mammals grow and flourish.

LONDON CLAY FOSSILS

London is mainly built on a wide basin, which in the early Eocene was filled by a great deposit of clay, up to 170 metres (550 feet) thick. In some areas, such as the Isle of Sheppey, this clay contains fossil seeds and fruits belonging to nearly 100 different genera. Some of these are beautifully preserved as uncrushed three-dimensional specimens. Most belong to such plants as the palm-tree, cinnamon, magnolia and sequoia, whose living relatives are found in the tropics of Malaya and Indonesia. The fossil fauna, too, shows that the region must have had a tropical climate, for it includes crocodiles and turtles.

Two remarkably well preserved specimens from the Isle of Sheppey clay.

Every feature of the modern landscape has been created during the 70-million-year span of the most recent era of the earth's history. Mountains created in earlier times, such as the Rockies, have been worn down and uplifted to their present heights. New ranges, such as the Alps and Himalayas, have risen from the sea. The continents have drifted thousands of miles to their present positions. The Pleistocene ice sheets have left their stamp over large tracts of the northern continents. The combined events of the Cenozoic Era have created a world in which there are, perhaps, greater differences in relief and climate than have ever existed before. These differences have had a considerable effect on the evolution of plants and animals.

The Mesozoic world had provided a fairly uniform environment for life forms. For the most part there had been one single land mass populated everywhere by similar plants and animals. But towards the end of the era the supercontinent had begun to break up and the individual pieces to drift apart. Now, in the Cenozoic, nature was able to start afresh, building in the shifting, separating continents new communities of both plants and animals.

The Age of Mammals

Great changes in animal life mark the opening of the Cenozoic Era. Apart from buried bones, imperceptibly changing into the hard fossils we find today, no trace remained of the dinosaurs or of many other Mesozoic animals. The little mammals which had survived in obscurity alongside the dinosaurs now took advantage of their new opportunities. But it was several million years before the slow process of evolution was able to fill all the niches left by the dinosaurs, especially the large herbivorous species.

As new groups of animals evolved in different areas, they found seas barring their passage to many parts of the world. The separate continents therefore came to contain quite different groups of mammals. Consequently, the variety of Cenozoic mammals was

THE MEDITERRANEAN DESERT

The Mediterranean Sea loses more water by evaporation each year than it receives from the surrounding rivers. It remains a sea only because water enters the Mediterranean basin from the Atlantic through the Straits of Gibraltar. If these were closed, the Mediterranean would be dry within a thousand years. This may in fact have happened some six million years ago, in the late Miocene. Buried below the present sea bed, there is a layer of rock salt two to three kilometres (one and a half to two miles) thick. When this was laid down, the Mediterranean must have been a vast sunken desert. The Nile and Rhône carved great valleys as they cut down to the level of this broad basin, within which their waters gradually shrank and petered out. The Mediterranean refilled and dried out again several times before it was permanently inundated at the beginning of the Pliocene.

Right: In the late Cretaceous, placental mammals (P) evolved in Asia and spread to western North America; some travelled on into South America. Marsupials (M) may have evolved in South America, and from there colonized North America and probably also migrated into still-joined Antarctica and Australia.
In the early Cenozoic, North America was the centre of placental evolution, and marsupials (m) became less important. From there, placentals spread into Asia, Europe and Africa. Marsupials were probably present in Antarctica and Australia. Later in the Cenozoic, monkeys and rodents arrived in South America, but there is still much argument as to whether they travelled there from North America or from Africa.

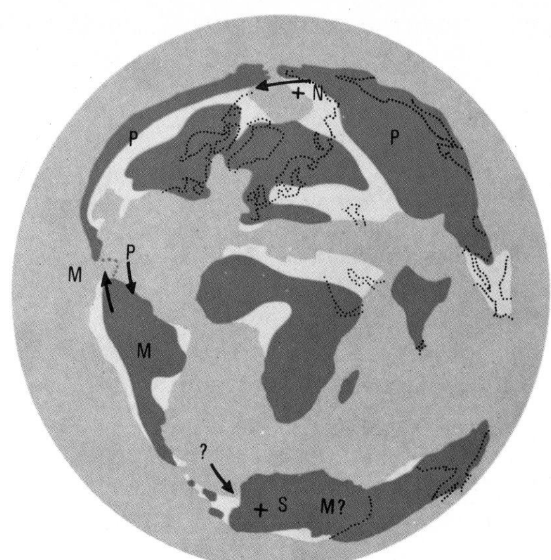

LATE CRETACEOUS

EARLY CENOZOIC

much greater than the variety of Mesozoic reptiles which they replaced. Today that variety has been greatly reduced by two agents – ice and man.

The Flowering Plants Flourish

Unlike the mammals, the flowering plants had already started to diversify into different families by early Cretaceous times – before continental drift had broken up the supercontinent into individual land masses. Naturally, some families appeared later within a particular continent, such as the cactus family in South America. But, in general, the flowering plant floras are very much alike throughout the world, with such families as the grasses and Compositae conspicuous everywhere.

In early Cenozoic times, the climates of the world were still warm and mild, so that fairly rich vegetation grew as far north as Greenland and the Canadian Arctic. Later, as the temperatures fell, distance from the equator became a crucial factor and narrow climatic bands developed. It was these climatic differences, not different evolutionary origins, that created the diverse floras which exist today, from the bare Arctic tundra to the luxuriant tropical rain forests.

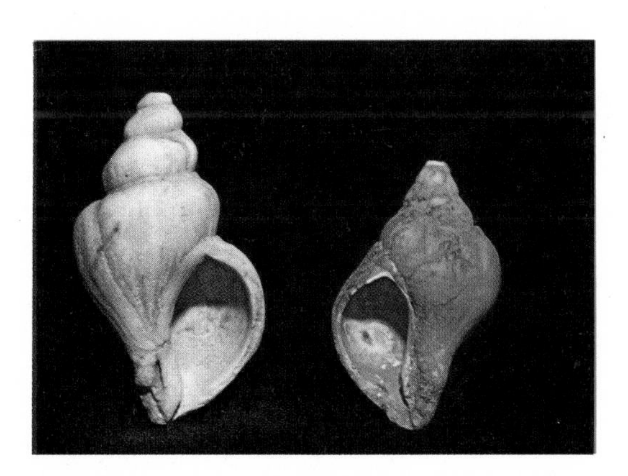

Left: Shells reveal climates. The left-handed whelk (Neptunea contraria) on the right is a warm-water species. Fossils are common in the Lower Pliocene crags (see page 20), but become rare in the Upper crags and are replaced by the common whelk (left), showing that the climate was cooling down at the end of the Pliocene.

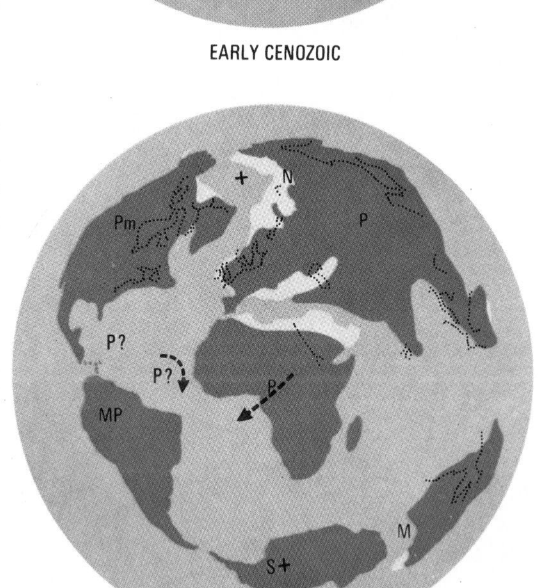

LATER CENOZOIC

Above: A reconstruction of Megazostrodon, a shrew-sized mammal from the late Triassic.

The Rise of the Mammals

Below: Oligokyphus, one of the last of the mammal-like reptiles, represents a half-way stage in the evolution of mammals. Though strictly a reptile, it probably had a scanty covering of hair and could grind its food. A false second palate, which allowed breathing to continue while the animal ate, is another typically mammalian feature.

One of the problems facing every animal is that of keeping warm – or, at least, maintaining a constant body temperature. And one of the solutions is sheer bulk. As an animal grows larger, the surface area from which heat is lost becomes relatively smaller. Evolution in this direction produced the huge but uninsulated dinosaurs of the Mesozoic Era, dependent on equable temperatures and doomed to extinction as a result of climatic changes.

The mammals evolved another solution to the problem, and one that was not dependent on size. The earliest mammals, such as the late Triassic *Megazostrodon*, were only the size of a mouse. But they had developed a hairy covering which would keep the heat loss from even a small animal down to manageable levels.

Warm and Safe

The problem of keeping warm is not, however, confined to adult animals. Eggs and embryos must also be protected from the cold. Birds solved this problem by incubating their eggs with their own body heat. The mammals solved it in three different ways. The most primitive group, the monotremes, lay eggs like birds, and keep them warm in burrows. The other two kinds, the placentals and the marsupials, developed even more efficient methods.

In the placental mammals, the eggs are retained inside the body of the female. After fertilization, the developing embryos remain warm and safe until they are ready to be born. Food and oxygen are continually supplied to the embryos and their waste products are carried away. The exchange takes place by diffusion between the blood of the embryos and that of the mother. The two blood systems are separated from one another only by a few thin layers of tissue which make up the placenta (hence the name 'placental mammals'). This system is so effective that a young placental mammal can spend many months developing inside its mother before it is born.

Most marsupials have no placenta and so give birth to their young at a far earlier stage than the placentals. The newborn marsupial is a tiny creature, no bigger than a kidney bean and quite incapable of self-support. Immediately it is born, therefore, it crawls into the protection of its mother's pouch. The pouch is a flap of skin concealing a number of teats. The embryo attaches itself to one of these and is nurtured by its mother for several weeks.

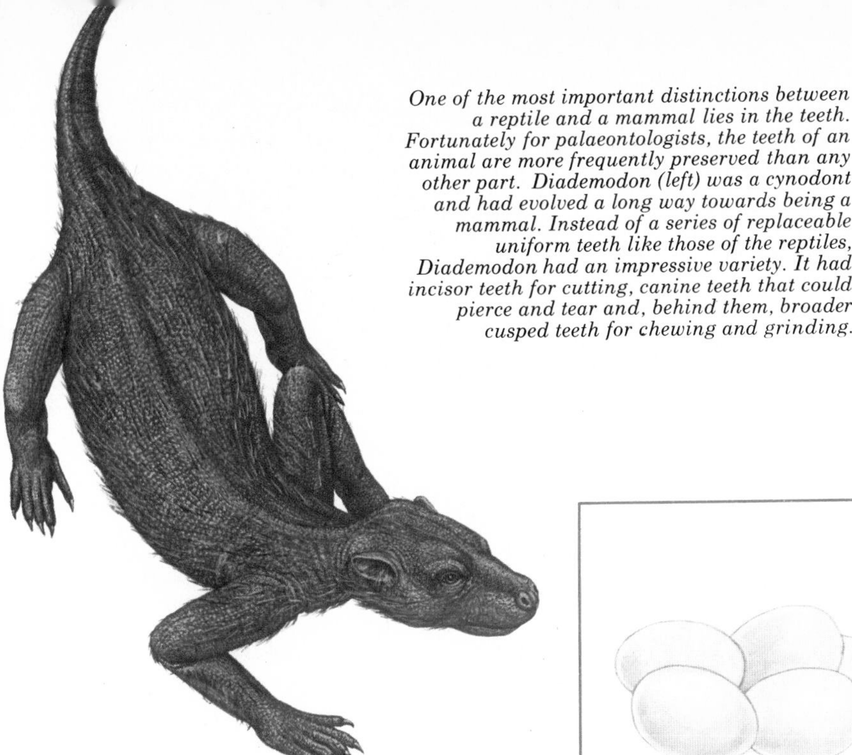

One of the most important distinctions between a reptile and a mammal lies in the teeth. Fortunately for palaeontologists, the teeth of an animal are more frequently preserved than any other part. Diademodon (left) was a cynodont and had evolved a long way towards being a mammal. Instead of a series of replaceable uniform teeth like those of the reptiles, Diademodon had an impressive variety. It had incisor teeth for cutting, canine teeth that could pierce and tear and, behind them, broader cusped teeth for chewing and grinding.

Lessons to be Learned

Just as unguarded eggs are an invitation to predators, so are unguarded young. When it breaks out of its egg, the young reptile is usually alone in the world. It has nobody to protect it and no parental behaviour to imitate. The young mammal, by contrast, spends most of its early life in the company of its parents. From them, it can learn about many of the dangers and opportunities which will confront it as an adult. The mammals' ability to learn is reflected in the enlarged size of their brains. The reptiles have relatively smaller brains: their patterns of behaviour often seem quite complex, but in reality they are merely instinctive reactions, fixed and unvarying, as though performed by elaborately programmed robots.

Non-Stop Breathing

Animals with 'warm blood' are far more active than cold-blooded animals. They need more oxygen and more food. Accordingly, one important difference between reptiles and mammals is that the latter have diaphragms. The diaphragm is a muscular partition between the lungs and the abdominal cavity. It helps to force air in and out of the lungs and enables more oxygen to be taken in.

All-Purpose Teeth

Reptiles swallow food whole and digest it while at rest. Their teeth are simple conical structures which are continually being shed and replaced. A mammal grows only two generations of teeth during its lifetime. The set which replaces the 'milk' teeth is permanent, and so the upper and lower set can form a permanent relationship. Projecting cusps on one set fit into depressions in the other, and sharp crests can shear past one another. The different types of tooth enable the mammals to seize their food and chew it more efficiently.

REPTILE v MAMMAL

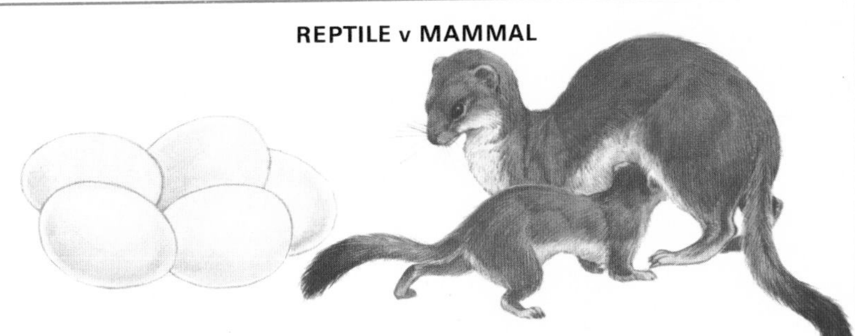

Instead of laying fragile, vulnerable eggs, most mammals give birth to live young. Unlike reptiles, they also take care of their offspring and feed them on milk from the mother's body.

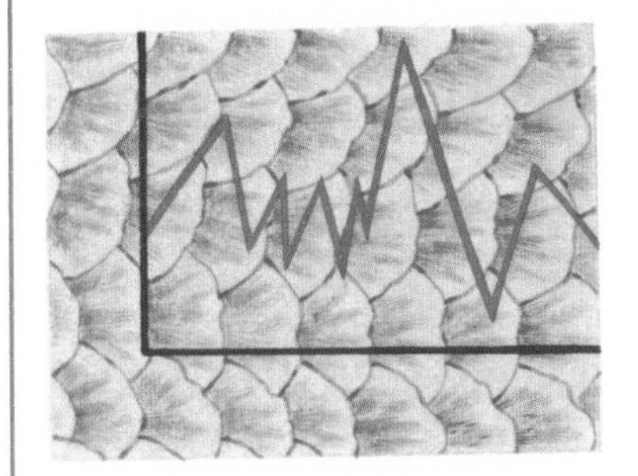

A covering of hair helps the mammals to maintain a constant temperature, while the scales of the reptiles provide no insulation, thus condemning the animals to the hazards of the elements.

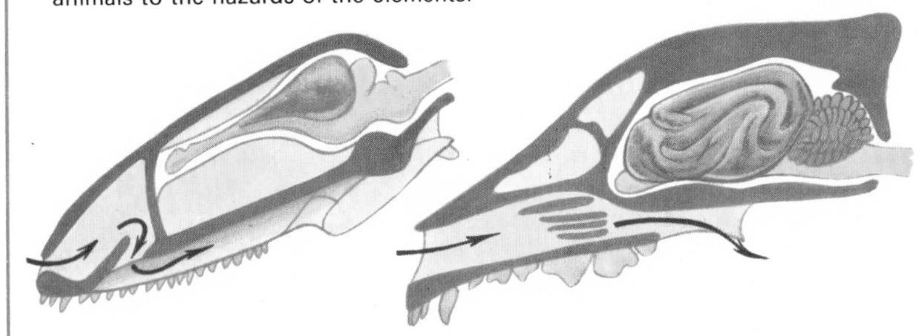

In a reptile, air must pass through the mouth, so that it cannot breathe while feeding. In mammals the air passes direct to the back of the mouth, and is drawn into the lungs by the action of the diaphragm. The cerebral hemispheres (grey areas) of the mammals' brain are also strikingly larger.

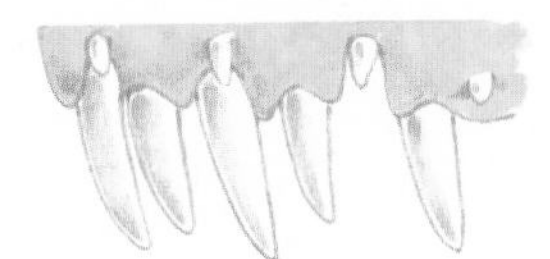

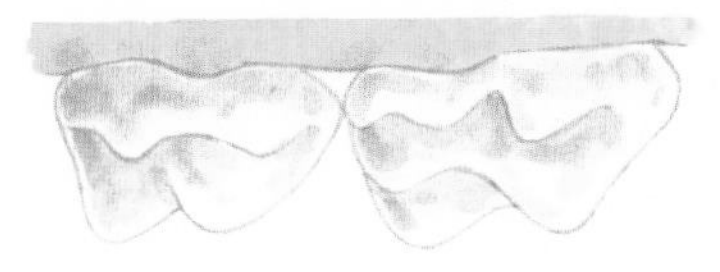

The many generations of reptilian teeth are useful only for gathering food, whereas mammals have complex cusps for crushing their food.

The Mammals of the Mesozoic

The tiny creatures that scampered warily about the feet of the great dinosaurs were so like their ancestors, the synapsids (mammal-like reptiles) that, with only a handful of fossil jaws, teeth and skull bones available, early palaeontologists found it difficult to distinguish between them. More recent finds, however, reveal that even during the Triassic and Jurassic Periods, the true mammals had become so abundant that they had spread to all parts of the world except Australia and South America.

Furtive Feeders

The early mammals were furry, shrew-like egg-layers, no more than a few centimetres long. As far as we can tell, they were nocturnal creatures which hid from the marauding dinosaurs by day, and ventured out to forage only at night, while the great brutes slept.

One early mammal was *Morganucodon*, specimens of which have been found in sites as far apart as China and Great Britain. This little creature probably behaved like a modern insectivore, feeding on worms, insects, grubs and the eggs and young of the smaller reptiles and amphibians. Such a diet would have provided the high-protein food which it needed to fuel its small, active body.

A Diet of Fruit

One group of Mesozoic mammals – the multituberculates – developed a new feeding habit. They evolved in the late Cretaceous, when flowering plants were first becoming common. Instead of keeping to a diet of flesh, they took advantage of this new source of protein. Creatures such as *Ctenacodon*, *Taeniolabis* and *Ptilodus* were all adapted to a diet of nuts and fruit. Like modern rodents, they had long, gnawing incisor teeth and large, complicated cheek teeth for chewing.

Death and Descendants

Most of the early mammals were doomed to become extinct before they had had a chance to take over from the dinosaurs. But it was not the reptiles which wiped out the Mesozoic egg-laying mammals: it was the physiologically superior descendants of the mammals themselves. The two more advanced mammalian groups – the placentals and the marsupials – were probably both descended from the pantotheres, a Jurassic group which included *Amphitherium*. Another Jurassic group, the docodonts, probably gave rise to the tenacious modern egg-layers – the monotremes.

Above: Taeniolabis was a Cretaceous multituberculate. These rodent-like animals fed on fruit and nuts. They became extinct in the Eocene, probably as a result of competition from the placental rodents.

The Mammals Take Over

As soon as the dinosaurs had disappeared, the new mammals hastened from their ecological obscurity to fill all the vacant niches. Their early evolution can be traced from fossils discovered in western North America, in sediments laid down by the erosion of the rising Rocky Mountains. Early Paleocene faunas included not only small insectivores and multituberculates like *Taeniolabis*, but also opossum-like marsupials and primitive placentals, both carnivores and herbivores. Throughout the Paleocene Epoch

Below: Modern representatives of the three kinds of mammal. The duck-billed platypus (left) is a monotreme, laying its eggs in a burrow near the river where it feeds. The koalas (centre) are marsupials. The cubs emerge from the pouch after about six months and are then carried on the mother's back. The mice (right) are among the most successful placentals. They have spread to all parts of the world and adapted to living as scavengers in urban areas.

these creatures grew larger and larger. By the end of the Paleocene there were carnivores (creodonts) the size of wolves and herbivores (condylarths) the size of pigs. Some of the insectivorous placentals had taken to the trees and adopted a diet of leaves and fruit. These became the earliest primates, from which the monkeys, apes and man himself eventually evolved – though, as we shall see, most of their later evolution took place in Africa.

Early Ungulates

The first herbivorous placentals were browsing animals which lived in the vast forests of North America. *Phenacodus*, a condylarth, had a long body, carried on short, sturdy limbs which were rounded off by small hooves. The hoofed mammals, or ungulates, became heavier and heavier. Apart from the condylarths, two groups were represented in the Eocene – the dinoceratids and the pantodonts. The latter included *Pantolambda*, which was the size of a sheep, and *Barylambda*, which was as big as a donkey. *Coryphodon*, also a pantodont, was even bigger, with broad, spreading feet and tusk-like canine teeth. *Uintatherium*, a dinoceratid, was as big and ugly as a rhino.

The History of the Horse

Early in the Eocene Epoch a new species of herbivore emerged. This was *Hyraco-*

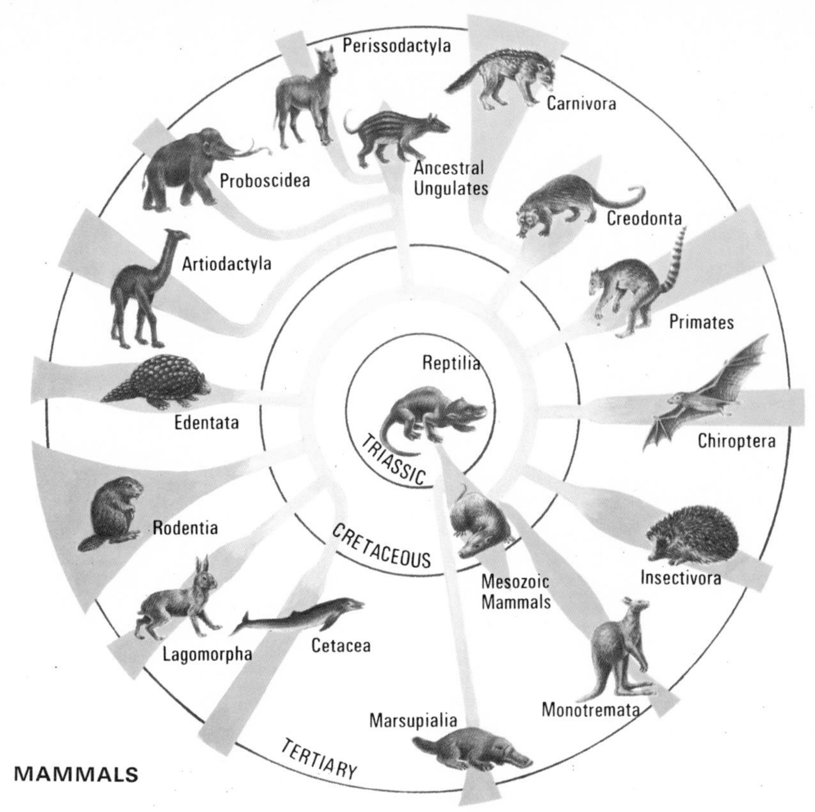

MAMMALS

The chart shows the evolution and probable relationships of the mammals. The widths of the branches show the relative periods of success and decline. Each branch has one example, as follows:

REPTILIA *Chiniquodon* (a possible ancestor)
MESOZOIC MAMMALS *Morganucodon*
MONOTREMATA *Ornithorhynchus*
MARSUPIALIA *Sthenurus*
INSECTIVORA *Erinaceus*
CHIROPTERA *Palaeochiropteryx*
PRIMATES *Notharctus*
CREODONTA *Oxyaena*
CARNIVORA *Hyaena*
ANCESTRAL UNGULATES *Phenacodus*
PERISSODACTYLA *Hyracotherium*
PROBOSCIDEA *Parelephas*
ARTIODACTYLA *Alticamelus*
EDENTATA *Glyptodon*
RODENTIA *Castoroides*
LAGOMORPHA *Eurymylus*
CETACEA *Basilosaurus*

Left: Uintatherium, a grotesque rhino-sized herbivore from the Eocene. It had tusky teeth and a series of knobs and horns on its head which grew proportionately larger as the animal grew older.

Right: Coryphodon was a pantodont, about the size of a donkey. Living in the late Eocene, it was the last North American pantodont, but its relatives survived in Asia until the middle Oligocene.

therium (*Eohippus*), the first in a series of forms leading to the modern horse, *Equus*. Each stage in the evolution of this line can be clearly traced in the fossil record. *Hyracotherium* was about the size of a fox. It was well adapted to the life of a forest browser, with spreading toes for picking its way through the damp forests. As the climate grew cooler and drier, grassland spread across North America. *Hyracotherium* changed its physique to suit its changing environment.

The descendants of *Hyracotherium*, including *Orohippus*, *Mesohippus* and *Miohippus*, gradually grew bigger. Their legs grew longer and their feet grew stronger. By the end of the Oligocene they could run fast enough to live in open plains, where flight, not camouflage, would be their best defence. The horses had to adapt not only to running on the plains, but also to an unvarying diet of grass. The cell-walls of grass are so hard that they would quickly have worn away a browser's teeth. As a result the grazing horses evolved longer, deeper teeth. The first true grazer was *Merychippus*, which evolved in the late Miocene.

LEFT BEHIND

Surviving almost unchanged since the Cretaceous Period, the American opossum is a 'living fossil' and the only marsupial in North America. A small catlike animal, it climbs through the trees using its prehensile tail as a fifth limb.

Once they had moved to the plains, the horses spread in all directions. From North America, *Hipparion* migrated to all the continents except South America and Australia. *Pliohippus* migrated to South America and gave rise to *Hippidium*. Another of its descendants, *Equus*, evolved in North America and migrated to the Old World. Here the genus still survives in the shape of horses, zebras and donkeys.

Odd-Toes

The horses belong to a group of ungulates called perissodactyls. The word 'perissodactyl' means 'having an odd number of toes'. The members of this order thus have either three hoofs on each limb or a single large hoof. Early Cenozoic North America was also the home of two other groups of perissodactyls – the brontotheres, or titanotheres, and the chalicotheres. Like the horses, the brontotheres had evolved from small herbivores in the early Eocene. By the middle of the Oligocene they had turned into lumbering giants and had spread into Asia and Europe.

Early brontotheres such as *Lambdotherium* and *Manteoceras* were harmless, but later forms such as *Brontops* and *Megacerops* sported large and dangerous horns on their noses. *Brontotherium*, the biggest of the group, stood two and a half metres (eight feet) high at the shoulders and had a V-shaped horn. Unlike the horses, the brontotheres were

Little Hyracotherium is dwarfed by its larger three-toed descendant, Hypohippus.

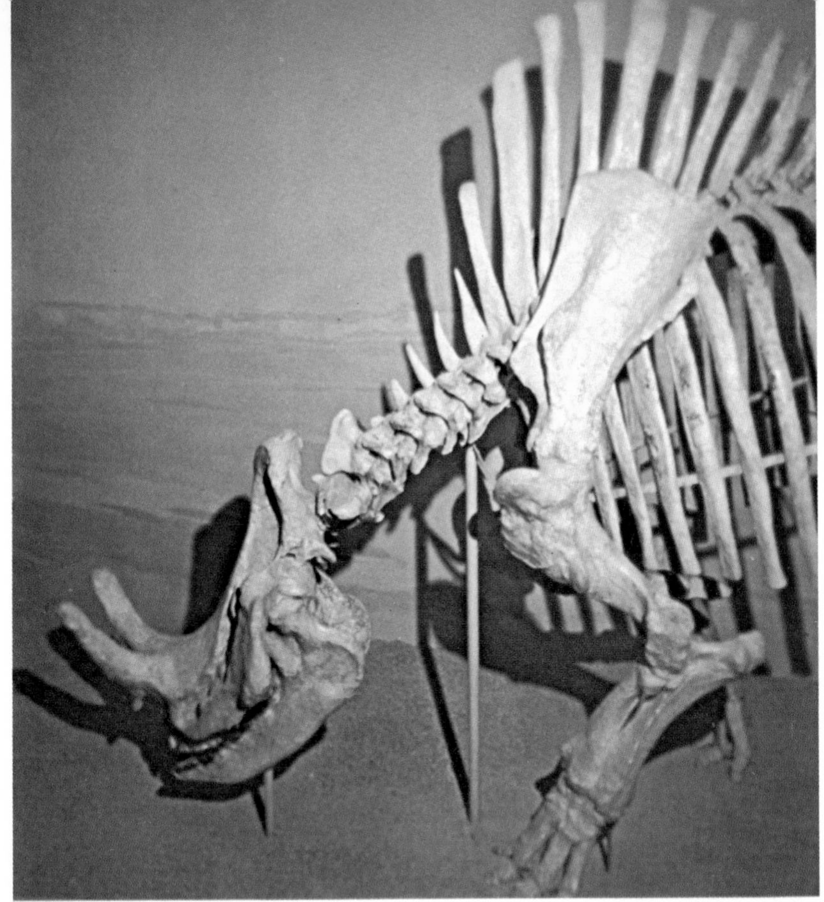

minimum of time feeding on the grasslands at the mercy of hungry predators.

The various kinds of artiodactyl have been very successful both in the New World and in the Old. The African fauna includes a vast array of antelope species; the Eurasian fauna is rich in deer. In North America, the massive bison and the pronghorn still bear witness to the flowering of this diverse group.

The camels, too, diversified greatly during the Cenozoic, though they had not yet spread beyond North America. The group embraced a great variety of species. In the Miocene Period these ranged from the graceful *Stenomylus*, which resembled a gazelle, to the lanky *Alticamelus* and *Oxydactylus* with their long, giraffe-like necks and legs.

Above: A fossil of Megacerops, one of the huge brontotheres that took the place of the early ungulates on the grasslands of North America.

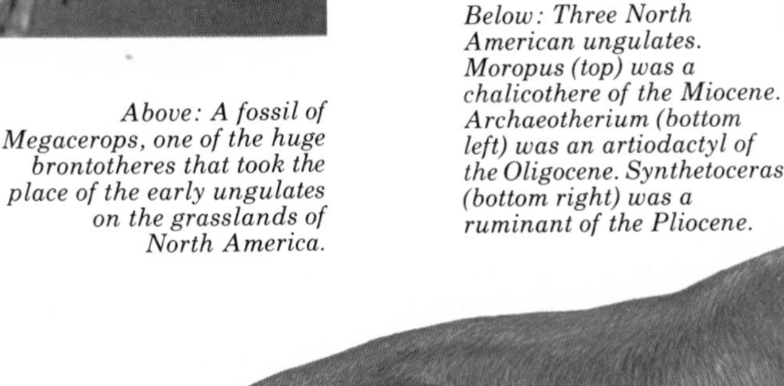

Below: Three North American ungulates. Moropus (top) was a chalicothere of the Miocene. Archaeotherium (bottom left) was an artiodactyl of the Oligocene. Synthetoceras (bottom right) was a ruminant of the Pliocene.

short-lived. By the end of the Oligocene they had become extinct.

The chalicotheres, such as *Moropus* (a horse-like form from North America) and *Macrotherium* (from Eurasia) never reached such huge proportions. They had claws on their feet instead of toes; these were probably used for digging up roots and tubers. The chalicotheres died out soon after the brontotheres. Among the perissodactyls, only the horses, tapirs and rhinoceroses survived.

Even-Toes

The decline of the perissodactyls was accompanied by the rise of the artiodactyls – ungulates with an even number of toes. Today this order contains a vast number of groups, including camels, giraffes, antelopes, deer, hippopotamuses, sheep, pigs, goats and cattle. The first artiodactyls were small, pig-like creatures which lived during the Eocene. *Archaeotherium* lived during the Oligocene Epoch. Trundling about like a cow-sized pig, it used its heavy canine teeth to grub up roots. Later artiodactyls developed a highly specialized technique for digesting plant food – the ability to chew the cud. This enabled them to spend a

The South American Ark

Isolated for much of its Cenozoic history. South America was the home of a fascinating collection of marsupials and placentals. Because the latter were all herbivores, the marsupials were able to survive alongside them. It was not until near the end of the era that their peaceful seclusion was shattered by the invasion of the North American placentals (see panel on page 133).

The marsupials evolved into a variety of insectivorous and carnivorous forms; the carnivores preyed on placental mammals. The fiercest predators were the borhyaenids. *Borhyaena* was an aggressive, wolf-like animal which lived in the Miocene Epoch. Its canine teeth were long and sharp. During the Pliocene *Thylacosmilus*, a later borhyaenid, was the scourge of the herbivores. Like the placental sabre-toothed tiger, it had cruelly-curved incisor teeth for piercing the hide of its prey.

Peaceable Placentals

In the lush forests and grasslands of South America, the hoofed placentals radiated into a great variety of forms. Some were very similar to those found in other parts of the world, and provide perfect examples of 'convergent evolution' (see page 26). Three orders probably arose from the condylarths which reached South America in the Paleocene. These were the litopterns, the notoungulates and the astrapotheres. Like the horses, some of the litopterns lost their toes and evolved a single hoof. In *Thoatherium* the side digits were even more reduced than in *Equus*. *Macrauchenia*, another litoptern, was rather like a camel in bone structure, but probably had a small trunk-like extension of the nose.

The notoungulates were the most varied and numerous of the South American placentals. They ranged from hare-like creatures such as *Pachyrukhos* and *Protypotherium* to lumbering giants more like rhinoceroses. The toxodonts evolved into forms of increasing size, from *Thomashuxleya*, an animal the size of a sheep, to *Scarrittia*, the size of a pony, and finally to *Toxodon*, a ponderous beast the size of a hippopotamus. The astrapotheres were gigantic creatures almost from the earliest days of their evolution. *Astrapotherium*

Above: Pyrotherium, the South American equivalent of the elephant.

Left: Macrauchenia, a camel-like litoptern that managed to survive for some time after the invasion of the northern placentals.

Right: Megatherium, a gigantic ground sloth, was brought to extinction by the ruthless persecution of man – probably within the last few thousand years.

Above: The rhino-like Toxodon was the last of its line. Its fossils were discovered by Darwin when he visited South America.

Above: Thylacosmilus was the largest and fiercest marsupial ever known, but it could not compete with the carnivorous placentals from the north.

DEATH FROM THE NORTH

Flourishing in their island continent, the unique South American mammals remained undisturbed for most of the Cenozoic. But in the late Pliocene the movements of the continents upset their peaceful seclusion. For nearly all the marsupials and placentals, the establishment of a land link with North America via the Panama Isthmus spelt disaster.

North America had long been united with the other northern continents by way of the Bering land bridge to Asia. Its Pliocene inhabitants were the survivors from many years of competition between different mammalian groups which had evolved and spread throughout this great land mass. The marsupials and primitive placentals of South America stood little chance of competing successfully with the army of aggressive carnivores and adept herbivores that was marching relentlessly southwards. At first the advance was a mere infiltration of rodents and other small creatures, but soon camels and other specialized herbivores arrived to take over the feeding grounds of the South American ungulates. The marsupial carnivores, too, lost their pride of place as the fierce placental cats completed the invasion during the Pleistocene.

Not all the South American mammals were vanquished, however. *Glyptodon* and *Megatherium* even made an advance into North America in the face of the enemy. It was only the advent of man, the most deadly hunter of all, that finally led to their extinction. Opossums, armadillos and the South American porcupines also spread northwards and are found today as far north as Canada. In the rain forests of South America, too, some of the more specialized animals escaped the pressures of competition. Today a few marsupials survive alongside the placental anteaters, armadillos and tree sloths.

Left: Glyptodon, an ancient relative of the armadillos, ate almost any kind of food it could find on the ground.

was more than one and a half metres (five feet) high at the shoulders. *Pyrotherium*, another (unrelated) herbivore, was even bigger. About three and a half metres (12 feet) long, it was as large as an elephant – and it possessed a trunk.

Of the South American hoofed mammals, only the notoungulates and the litopterns survived into the Pleistocene Epoch, and of these only big creatures such as *Macrauchenia* and *Toxodon* managed to hold out for any length of time against the northern invaders.

Ground Sloths and Glyptodonts

Not all the South American placentals became extinct. Descendants of one group, the Edentata, survive to this day. One of the earliest edentates was *Metacheiromys*, a little creature like the modern armadillo but without its armour plating. It was not long, however, before its descendants developed their protective coats – tough bony plates arranged in articulated bands which enabled the animals to curl up when they were in danger. The edentates were related to the glyptodonts, which were rather like mammalian tortoises. *Glyptodon*, a massive tank of a creature, had a solid dome of armour on its back and a covering of bony plates over its head and tail.

Late in the Cenozoic, a group of unarmed but even more spectacular edentates evolved. These were the ground sloths. Related to modern tree sloths, they were almost six metres (20 feet) long. Like elephants, they browsed on the leaves of trees. Resting on their haunches and contorted feet, they used their clawed hands to pull branches down to their mouths. These magnificent but awkward animals were probably wiped out by early man, who found them easy to outwit.

Splendid Isolation

Cast adrift from the mainland long before the placentals could jump aboard, the Australian 'life raft' has enabled a unique collection of primitive mammals to survive. Apart from rats, bats, and a few placentals introduced by man, such as rabbits and dingoes (wild dogs), the only mammals found in Australia and New Guinea are marsupials and monotremes.

The Mystery of the Monotremes

The fossil record of the monotremes is so scanty that their origins are quite unknown. None of the usual clues is available. The two living forms lack proper teeth, so that even their relationships with other mammals can be no more than guessed at.

Both the spiny anteater (*Echidna*) and the duck-billed platypus (*Ornithorhynchus*) are highly specialized in their feeding habits. The platypus is a semi-aquatic creature, with webbed feet for swimming and a flat duck-like beak for scooping up worms and other small creatures from the river bed. It also uses its beak to dig the burrow in which it lays its eggs. The spiny anteater looks rather like a hedgehog with a long snout. It has powerful clawed feet which are used to break open ants' nests.

The Marsupials

Our record of Australian marsupials does not begin until the early Miocene. At that time Australia had a tropical or subtropical

Above: Dasyurus, commonly known as a 'native cat', is one of the few remaining marsupial meat-eaters. It preys on other small marsupials.

Below: A scene from Pleistocene Australia. On the right are two enormous wombat-like Diprotodon. In the centre, the giant kangaroo, Procoptodon, nurses its growing offspring. A 'mob' of these creatures bounds away in the background, on their huge back legs. When kangaroos are first born, the front legs are bigger than the back legs because they have to drag themselves up to the mother's pouch. On the left is Thylacoleo, the marsupial 'lion'. Palaeontologists are puzzled by its teeth. At the front it had small incisors and canines, while its cheek teeth were long and shearing. It has been suggested that Thylacoleo was a carnivore, but it seems more likely that the cheek teeth were used to slice up fruit or other vegetable matter.

NEW ZEALAND

The fauna of New Zealand suggests that that region must have been isolated even longer than Australia, for it has no land mammals of any kind – neither monotremes, nor marsupials, nor placentals. Continental drift data show that it may have separated from Antarctica and Australia in the early Cretaceous, before these mammals had evolved – though the absence of the older Mesozoic mammals is a puzzle. The lack of competition from other land vertebrates provided an unusual opportunity for the flightless birds. The kiwi survives today, and until a few hundred years ago there were many different species of moa. These are ratite birds, related to the flightless ratites found in other southern continents. Their ancestors must have lost the power of flight in adapting to ground life.

climate which provided favourable conditions for the fossilization of animals. The interior of the continent was forested, and contained numerous lakes and swamps. Nearly all the fossils found in Australia belong to groups which are still in existence; among the animals represented are the marsupial wolf, cat, bandicoot, phalanger, wombat and kangaroo. There are three groups of marsupials living today – the dasyurids, the parameloids and the diprotodonts.

The only carnivores on the continent are the dasyurids. These include *Thylacinus*, a dog-like animal, commonly known as the Tasmanian wolf. It is almost identical to the placental wolf and to the extinct marsupial borhyaenids of South America. Other, smaller carnivores include the cat-like *Dasyurus* and *Sarcophilus*, the the 'Tasmanian devil'. These aggressive creatures prey on the rabbit-like bandicoots (parameloids) and the smaller diprotodonts. The diprotodonts are a diverse and unusual group of herbivores. Among modern forms are the kangaroos, wallabies, wombats, phalangers and the famous koalas.

Many modern diprotodonts are large, but their Pleistocene relatives were much larger. *Diprotodon* was a lumbering herbivore up to four metres (13 feet) long. Numerous specimens of this great beast have been found in the mud of Lake Callabonna in South Australia, some so well preserved that the contents of their stomachs could be identified. In the dry season, many plants grew on the salty crust which formed around the edges of the lakes. It seems that *Diprotodon* fed on these plants, and that from time to time an unfortunate animal crashed through the surface crust and drowned in the mud below.

During the Pliocene and Pleistocene Epochs the Australian climate was gradually changing. The lakes and forests which had previously covered the centre of the continent were disappearing. In their place came grassland and desert. And to graze on the grass there arose a variety of marsupial forms which were equivalent to the placental cattle, deer and horses – the kangaroos and wallabies. These mostly grazed in 'mobs', but some were adapted to a browsing life. *Procoptodon* and *Sthenurus*, now extinct, were giants – more than two metres (six and a half feet) high.

> **SAFETY FIRST**
>
> The array of tusks, antlers and horns sported by so many herbivores looks at first sight like a vast armoury: it seems obvious that these aggressive instruments must be weapons of war. But in fact they are used mostly for mock fighting and trials of strength, and never with murderous intent. Often they are found only in the males, which suggests that they are more important in courtship contests and territorial battles than for use against predators. In many cases the weapons look fierce but would be incapable of inflicting damage on an adversary. Usually, the combatants can ram each other or wrestle together interlocked, but cannot pierce or slash each other's flesh. This is obviously an advantage: it would not be in the interests of any species if its members could kill or maim each other.

The Mammals of Africa

Today, deserts isolate Africa from the rest of the world almost as effectively as the oceans isolate Australia. Earlier in the Cenozoic, these deserts were covered with forests, and many types of mammal were able to roam freely between Africa, India and South-East Asia. But, just as it ended in isolation, so the Cenozoic history of Africa began with isolation, for shallow seas covered the northern lowlands of the continent.

The inaccessibility of Africa had a far-reaching effect on the country's early mammal fauna. Only a few placentals managed to cross the northern seas, and these failed to give rise to many new types. Only later, in the Miocene Epoch – when many other types of mammal migrated into Africa – did a fully diverse mammal fauna develop there.

The Elephants

Today there remain only two species of elephant – one in Africa and one in Asia – and even these are in danger of extinction. During the Cenozoic, however, they were a very diverse group which spread to almost every corner of the world. The earliest members of this order – the Proboscidea – lived in North Africa in late Eocene times. *Moeritherium* was a low-slung animal about the size of a big pig, with broad feet and hooved toes. It had neither tusks nor trunk. Later in the Miocene, the deinotheres emerged in Africa and gradually spread to Eurasia. These animals had tusks and trunks and were very large. *Deinotherium* was more than three metres (ten feet) high at the shoulder. Its tusks emerged from its lower jaw and curved back below the throat. What role these played it is difficult to imagine. But if the animal dropped to its knees they could have been usefully employed rooting up plants from the forest floor.

The group that gave rise directly to the elephants grew up in North Africa during the Oligocene. *Phiomia* was shaped like an elephant but its ears, trunk and tusks were relatively small. *Gomphotherium* (*Trilophodon*), its larger descendant, had

Above: Saghatherium, a relatively large hyrax that lived during the Oligocene Epoch.

Below: Moeritherium (left) was the forefather of the elephants. It had neither trunk nor tusks, although one pair of teeth were enlarged. Its descendant Deinotherium (right) had very advanced proboscidean features, but no intermediary forms have been found to show how the evolution took place.

developed a long trunk and two pairs of tusks. The successors of these animals grew to gigantic proportions. Some migrated to the Americas and some, such as *Platybelodon* (see page 138), to Eurasia. Two distinct groups emerged. One group which ranged over Eurasia, North America and Africa is typified by the Pleistocene *Mammut (Mastodon)*. The other comprises the mammoths and the modern elephants.

Cousins Large and Small

The elephants' only living relative, the hyrax *Procavia*, could hardly look less like its giant cousins. The hyraxes evolved during the Oligocene Period. Unobtrusive creatures, they looked and behaved like rabbits – but they were, in fact, ungulates, with tiny hoofs on their toes. Although small compared with the elephants, the hyraxes did grow fairly large. *Saghatherium* and *Megalohyrax* were about the size of a lion.

Arsinoitherium, an extinct relative of the elephants, was found in the Fayûm deposits of Egypt and named after the ancient Egyptian Queen Arsinoë. Nearly four metres (13 feet) long, it was the largest animal in the African Oligocene fauna. It looked rather like a rhinoceros, with great twin horns on its nose and a pair of smaller horns over its eyes. Unlike the horns of the rhino, however, which are formed of matted hair, those of *Arsinoitherium* had a bony core, like those of cattle. It was probably a forest-dweller, living on the foliage of trees. It did not survive into the Miocene Epoch.

The only other African mammals of the early Cenozoic were the sirenians (see panel) and the desmostylids (see page 139), both of which were aquatic creatures.

The Primates

Africa is also the continent within which we can see the most complete record of the evolution of the primates – the group which includes the little lorises and lemurs, the monkeys and apes, and man himself. The earliest, most primitive types of primate appeared first in North America and Europe, but the group became extinct in those areas at the end of the Eocene. Ancestors of both monkeys and apes were already present in Africa in the Oligocene Fayûm deposits, and it is even possible that the New World monkeys which

CARNIVORES

Dependent on food that is mobile, often armoured and sometimes very devious, carnivores must be alert, intelligent and strong to catch their prey. Herbivores obviously do not have such problems. A great variety of food is growing all around them, and many different kinds can find a niche in the same environment. For carnivorous animals, fewer different life styles are possible, and this may be why only two orders of placental carnivores – the Creodonta and the Carnivora – became adapted to a purely predatory life.

By the start of the Oligocene, the early predacious mammals, the creodonts, had almost died out. Their place was taken by the more highly specialized members of the Carnivora. This large order includes both marine mammals – the seals and sea lions – and land mammals – the cats, hyenas, dogs, bears, badgers, weasels, pandas and many more. Although not all equally advanced, these animals are mostly very intelligent, have well developed senses and strong teeth and limbs. Apart from man, they have few enemies. But today many are losing the battle against extinction as their habitats are destroyed.

As yet, neither ancestors nor descendants of Arsinoitherium have been found. It was a huge horned ungulate that appeared and disappeared abruptly in Lower Oligocene Egypt.

MERMAIDS

Ancient sailors, wearied by long and uneventful voyages, often told stories of the mermaids they had seen – strange beings, half woman and half beast. The only creatures they might have encountered that fitted this description were the sea cows which basked on the shores of the continents. These animals belonged to the sirenian order, of which the earliest fossil specimens come from the Eocene deposits of Africa. *Protosiren* of the Eocene and *Halitherium* of the Miocene were two early species. These hardly differed from the dugongs and manatees – great, lazy beasts some three metres (ten feet) long, which browse on the vegetation of rivers and coastal waters in the tropics today. The dugongs inhabit the Pacific and Indian oceans, while the manatees live on Atlantic coasts.

Halicore, the living sea-cow.

Right: Baluchitherium was the largest mammal ever to live on land. It was as tall as a giraffe but its neck was much thicker. An ancestor of modern rhinoceroses, it was unusual in not having a horn. But, with its great size, it would hardly have needed any other form of defence.

Above: Platybelodon lived in Asia during the Pliocene. The remarkable adaptation of its mouth and teeth must have given it an advantage over animals that had to cut or wrench the foliage from plants. Platybelodon could simply scoop up plants, roots and all, like a mechanical excavator.

appeared in South America in the Oligocene had somehow managed to get there from Africa across the South Atlantic, which was then narrower than it is today. Africa, too, has produced the most detailed record of the evolution of man himself. However, all the forest-dwellers of the late Cenozoic must have roamed throughout the tropical forest belt from Africa to South-East Asia, and they may be better known from Africa only because the increasing aridity and erosion of that continent have since made its rocks, and the fossils contained in them, more accessible.

As Africa became drier in the late Cenozoic, savannah grasslands spread across the continent, providing opportunities for the evolution of the many and varied herbivores which are so characteristic of Africa today – antelope, buck, eland, gazelle, kudu, oryx, wildebeest, giraffe, zebra and many others. These in their turn offered a plentiful supply of food for hyenas, lions and scavenging birds.

Mammals of Europe and Asia

Though Asia was the Cretaceous homeland of the placental mammals, much of their early Cenozoic evolution took place in North America and Europe, from which Asia was then separated. Later, in the middle Eocene, the Atlantic finally separated North America from Europe, which at the same time became united with Asia as the old Turgai Sea receded. Until the Oligocene, therefore, the faunas of all the northern continents were very similar to one another.

The Largest Land Mammals

One of the groups which became very diverse during the Cenozoic was that of the rhinoceroses. Some of the early rhinos were running forms with long, slender legs. Others, such as the North American *Teleoceras*, were semi-aquatic animals similar to hippos. The most stupendous of them all were the baluchitheres, such as *Indricotherium* and *Baluchitherium*. The latter was the largest land mammal ever. It roamed Central Asia in the Oligocene and early Miocene, browsing on tree tops like a super-giraffe. Its shoulders towered five metres (17 feet) above the ground and its long, muscular neck supported a mighty one-metre (three-foot) head. None of these giants has survived. When the shallow seas which had confined the

elephants to Africa receded in the early Miocene, these beasts spread into Eurasia. It seems that the giant rhinos could not compete with their more efficient feeding methods.

Back to the Sea

Like the reptiles, the mammals were not content merely to rule the land. They also invaded the seas in the shape of seals, walruses, sea cows, dolphins, porpoises and whales. Many gradually lost their fur coats and evolved thick linings of blubber as insulation. Their bodies became streamlined and their legs turned into flippers.

The whales are the most spectacular of the aquatic mammals. They swim by flexing their bodies up and down, rather than from side to side like fish. This action is assisted by the flukes which extend sideways from their tails. The blue whale is the largest animal which has ever lived; today it is fighting extinction in the southern seas. The whales first appeared during the Eocene Epoch. With the surrounding water to support their weight, they were soon able to grow very big. *Basilosaurus (Zeuglodon)* was a fish-eating whale with a slender, tapering body some 18 metres (60 feet) long.

The desmostylids were aquatic relatives of the elephants. *Palaeoparadoxia* lived on the shores of the North Pacific in the middle of the Tertiary Period. About two and a half metres (eight feet) long and heavily built, it used its strong crushing teeth for cracking the shells of the molluscs on which it fed. *Desmostylus*, a walrus-like creature, had a large head with projecting tusks which it may have used for prising molluscs from the rocks.

SEALS AND SEA LIONS

Around the coasts of the northern and southern oceans, the beaches are carpeted in spring with a writhing mass of enormous bodies as the seals, sea lions and walruses come ashore to breed. Though the sealions, walruses and fur-seals appear very similar to the true seals, the two groups each evolved independently from land carnivores. Even today, the true seals, unlike the others, cannot bring their hind limbs forwards below the body. The oldest fossils were found in early Miocene sediments in California, and the earliest known member of the group is *Enaliarctos*. Its skull was more than 20 centimetres (eight inches) long, and shows certain similarities to the skulls of early bears – the bears, too, being members of the order Carnivora. Modern bears still wade into the water to catch fish, and such a habit may have led eventually to the changes which took these animals back to the sea to find a permanent home there.

Below left: Although the early whales had not reached the immense size of some of today's whales, they were large animals. Basilosaurus (Zeuglodon) was 18 metres (60 feet) long. It had ragged spiky teeth for catching its prey of fish.

Below right: Desmostylus lived around the coasts of the northern Pacific in the Miocene Epoch. It must have lived in much the same way as a modern sea lion.

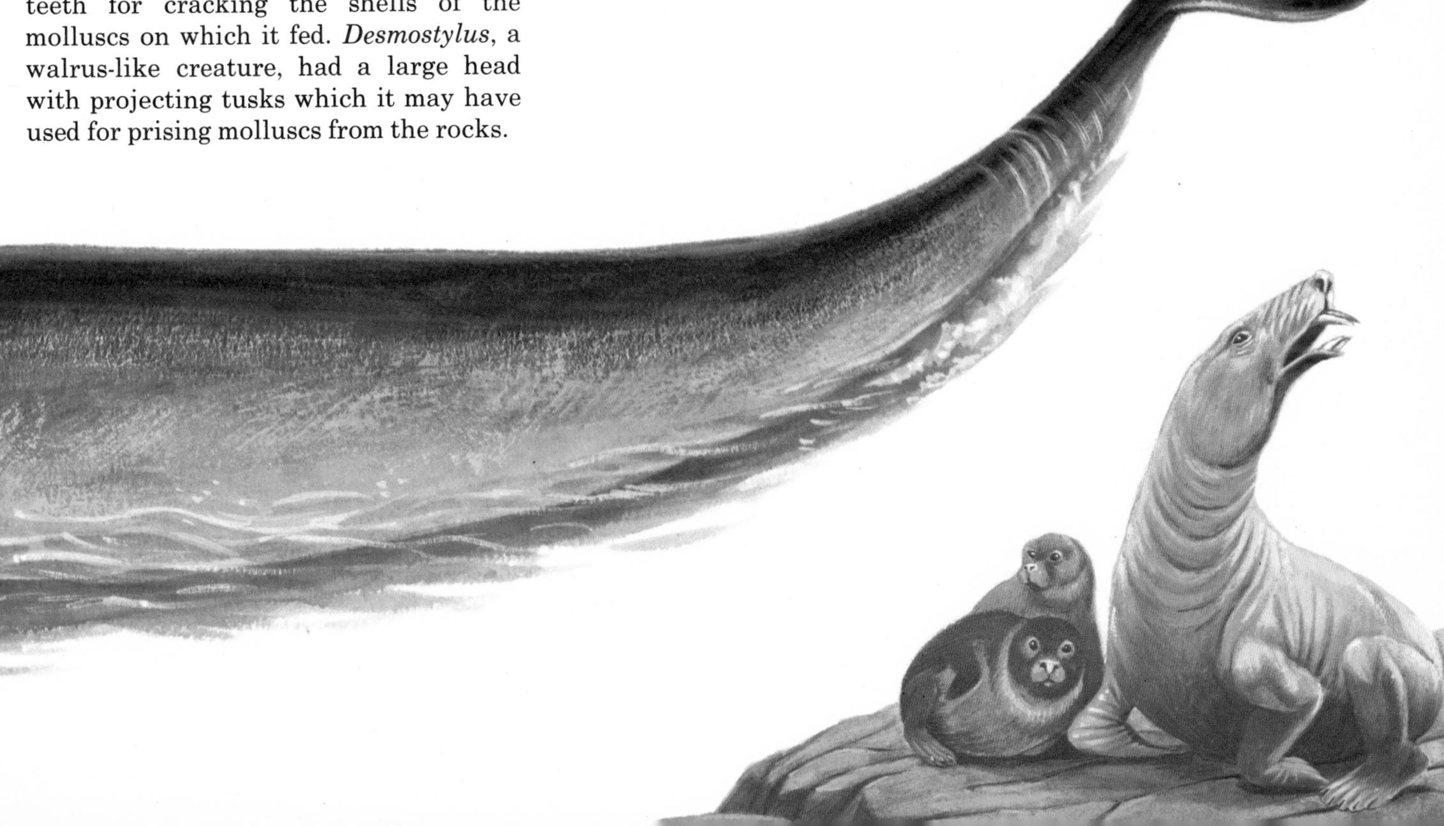

The Great Ice Age

The Pleistocene Epoch

For some 60 million years after the beginning of the Cenozoic Era, mammals continued to evolve and diversify throughout the world. Only Antarctica lay frozen and lifeless beneath a covering of ice that had begun to form early in the Cenozoic. A similar fate was later to befall the still warm and sunny lands to the north: the Ice Ages of the Pleistocene Epoch were approaching.

It was in the most northerly continents, Eurasia and North America, that the full effects of the climatic revolution were felt. Previously, in the late Pleistocene, much of these continents had been covered by savannah grasslands, over which had roamed a rich fauna of mammals. With the first breath of the approaching cold there came a cooler, more humid climate, and woodlands started to replace the grasslands. As the temperature dropped further, the winters became colder and longer, the summers cooler and shorter. The forests in their turn shrank, and were replaced by tundra. The first of the Pleistocene Ice Ages had arrived.

As the ice sheets advanced they scraped the land clear of soil and levelled forests in their path. They smoothed the tops of hills, scoured out valleys and carried huge boulders from their parent outcrops to distant resting places. When the glaciation was at its height, more than eight million square miles of North America, Europe and Asia lay shrouded in ice.

The effect of the Ice Ages on animal life was far-reaching. As the cold deepened, more and more of the warmth-loving mammals of the northern lands migrated south, some never to return. Other mammals remained in the north, gradually evolving in response to the colder conditions. Their living descendants include the brown, grizzly and polar bears, the elk, the wolf and the lynx.

Glacials and Interglacials

It is not known exactly when the Cenozoic Antarctic ice cap formed – an event which heralded the great climatic changes of the Pleistocene Epoch. Some scientists believe that it may have taken place 40 million years ago, but the first reliable evidence for it dates back only ten million years. It seems clear, however, that the climates of the world were cooling gradually throughout the Cenozoic Era and that the rate of cooling increased dramatically over the last few million years. This process culminated in a series of glacial periods, or Ice Ages, which affected the northern hemisphere in particular. At one time, scientists believed that the first of these Ice Ages started at the beginning of the Pleistocene Epoch, about 1·8 million years ago, but it is now clear that it did not begin until 1·5 million years ago.

In Europe, evidence for five glaciations has been found in the Alps. These were named the Donau, Günz, Mindel, Riss and Würm glaciations. The Ice Ages in Europe are named after these glacial periods, and corresponding glacial periods are known to have taken place simultaneously in North America. These were the Nebraskan, Kansan, Illinoian and Wisconsin glaciations – the American Wisconsin glaciation probably covering the time-span of both the Riss and the Würm glaciations of Europe.

A remnant of the Ice Age, this Swiss mountain glacier flows slowly down from the snow fields, but the melting of the lower reaches keeps pace with the flow, so the foot of the glacier remains stationary. A cooling of the climate, however, would slow down the melting, and the glacier would advance down the valley, perhaps joining with others on the way and covering the lowlands with ice again.

The vegetation was of a completely modern type, although it included many plants now found only in warmer regions. Neanderthal man hunted hippos and other large animals.

Horses roamed the grasslands, often pursued by cave lions, while beavers gnawed happily in the birch and aspen woods.

The huge straight-tusked elephant (Elephas namadicus) roamed the land, together with other large mammals. Homo erectus probably hunted these and supplemented his diet by collecting fruits.

This chart shows what life was like in western Europe during the Ice Ages and the interglacials. Obviously nothing could live in those areas completely covered by ice, but a surprising number of mammals managed to survive in the snowy wastes just to the south of the ice sheets, and man survived by hunting these animals. Conifers grew in many places, but the colder regions probably supported only lichens and grasses. The earliest (Donau) glaciation is not shown on this chart because as far as we know it affected only the Alpine regions. Fossils tell us that much the same kinds of plants and animals inhabited North America during the Ice Ages and interglacials.

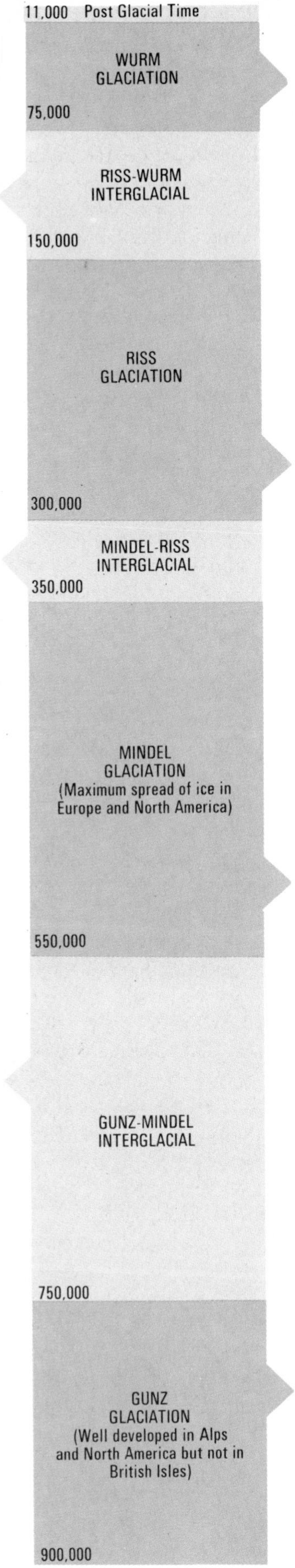

11,000 Post Glacial Time

WURM GLACIATION

75,000

RISS-WURM INTERGLACIAL

150,000

RISS GLACIATION

300,000

MINDEL-RISS INTERGLACIAL

350,000

MINDEL GLACIATION (Maximum spread of ice in Europe and North America)

550,000

GUNZ-MINDEL INTERGLACIAL

750,000

GUNZ GLACIATION (Well developed in Alps and North America but not in British Isles)

900,000

Conifers and dwarf willows dominate the sparse vegetation, while Cro-Magnon man and the arctic fox both find a source of food in the reindeer.

A woolly mammoth killed by the spears of Neanderthal man provided a wealth of meat for the human population, and some warm skins with which to make clothes.

A very cold period saw bleak scenes like those of the Arctic today. Musk oxen grazed on the sparse vegetation and were themselves attacked by wolves.

Wild boar and brown bears lived in the cold forests around the tundra. Pines and spruces made up most of the forest, just as they do in northern lands today.

Woolly mammoths, woolly rhinoceroses and well-furred reindeer all managed to find enough food on the fringes of the great ice sheets. Their thick coats kept them warm, and the mammoth is also thought to have kept a store of fat around its shoulders to help it through the winter. Although these mighty beasts managed to survive the harsh climate, they were no match for man, who finally hunted them to extinction about 10,000 years ago.

Each glacial period probably set in fairly gradually, over about 90,000 years, but ended more rapidly, over a mere 9000 years or so. Though each brought more severe climates to the lowlands, in the earliest two (the Donau or Nebraskan, and the Günz or Kansan) the glaciers were restricted to the mountains and did not descend to the lowlands. The Mindel or Illinoian glaciation seems to have been the longest and most severe, and the later glaciations seem to have been progressively shorter and milder.

Why Ice Ages Occur

The reason for these climatic variations may have been discovered by the Serbian physicist Milankovitch in 1913. He calculated the effects of regular variations in the angle of the earth's axis, and in the orbital path of the planet, on the duration and severity of the seasons. His results coincided remarkably well with the climatic fluctuations that occurred during the Pleistocene Ice Ages. (They also suggest that the glaciers will advance again within a few thousand years.)

Theories of this kind, however, explain only the comparatively minor fluctuations in climate which occur over a short period of time such as a few millenia. But the onset of an Ice Age climate involves a major alteration in the climatic régime of the earth, and such a change seems only to take place at very long intervals. Three hundred million years separate the present Ice Age from the Permo-Carboniferous Ice Age, and a similar period of time separated that Ice Age from an even earlier one.

Scientists are still uncertain of the reasons for these rare major changes. But it is perhaps not a coincidence that the Ice Age of 300 million years ago took place when the enormous land mass of Gondwanaland surrounded the South Pole, and that the Pleistocene Ice Ages similarly occurred when Eurasia and North America lay around the North Pole.

When the poles lie in open ocean, any snow that falls will usually melt in the sea water, and ice will not persist for long into the springtime. But when land surrounds the poles, the covering of winter snow reflects much of the sun's heat, so that less heat is absorbed by the ground. Winter cold may then extend later and later, until finally it lasts throughout the summer and an Ice Age begins. The presence of the small Arctic Ocean immediately around the North Pole may have delayed the onset of the Cenozoic Ice Age, and it is possible that this was finally triggered off by changes in water circulation in the Arctic Basin.

A glacier moving down a valley gouges rock from the sides and the bottom and produces a distinctive U-shape, such as can be seen in the valley above. The rocks and the finer materials are dropped further on when the glacier melts and they form deposits known as tillites, or boulder clays. By mapping the occurrence of the U-shaped valleys and boulder clays, we can discover just how far the glaciers spread in the Ice Age.

Adapting to the Cold

The onset of the Pleistocene Ice Ages compressed the climates towards the equator and radically changed the flora and fauna of the temperate lands. South of the great ice sheets which covered Canada and northern Eurasia lay a broad belt of treeless tundra. Further south again, patches of birch, poplar and oak forests started to appear, separated by grassy steppelands. Even as far south as Spain and Italy, the land lay covered by sombre pine forests.

The mammals that roamed these inhospitable wastes were very different from those which had lived there in warmer times. Some, such as the Ice Age mammoth and rhino, had evolved a warm covering of hair. For all the herbivores, food was in short supply, and they adapted to this problem by becoming smaller so that each individual required less food. For example, the woolly mammoths and brown

Above: A map of the northern hemisphere showing the maximum extent of the ice (shaded) during the Pleistocene Ice Age. The ice spread out from the north and from mountainous areas and covered half of North America, as well as most of the British Isles and large areas of Europe. Eastern Asia was much less affected, and this accounts partly for the much richer flora and fauna of that region when compared with those of similar latitudes in Europe.

bears of the glacial periods were about 20 per cent smaller than those of the interglacials. Predators, such as the cheetah, found that their prey provided fewer meals, and these too became smaller.

Evolutionary change took place rapidly in response to these new conditions, and most of the new Ice Age mammals had appeared by the time of the Mindel glaciation – the first to bring ice to the lowlands. Reindeer, musk-ox, the steppe mammoth *(Mammuthus trogontherii)* and the woolly rhino had all evolved by that time. Others appeared later, including cave bears, the woolly mammoth *(Mammuthus primigenius)* and the true elk or moose. The last, the polar bear, did not evolve until the Würm glaciation, only 50,000 years ago.

Frozen Solid

During Ice-Age spring-time, the streams carved deep channels into the soil under the still-frozen covering of ice and snow. Occasionally, the heavier mammals fell through this treacherous surface crust to their deaths, and their bodies were frozen and preserved indefinitely. At least 25,000 mammoths have been found buried in the Siberian permafrost, some preserved in almost perfect condition. They died over a period of many centuries, from over 44,000 to less than 11,500 years ago. Frozen woolly rhinos, bison and horses have also been found.

We know the appearance of the woolly mammoth very well, not only from these frozen specimens but also from the drawings and paintings of them with which the Ice Age men who hunted them decorated the walls of their caves.

Warmer Times

During the interglacial periods, when warmer climates returned, forests and a woodland fauna reappeared in Central Europe. The straight-tusked elephant, *Elephas namadicus*, then replaced the mammoth. Other newcomers were the woodland bison, the giant deer or Irish elk, and the large aurochs, the ancestors of modern

Below: The typical vegetation of the Arctic tundra today. This vast region around the Arctic Ocean is too cold for trees to grow, and the only plants are lichens, various herbaceous plants, and a few sprawling shrubs. Large areas of North America and Europe would have been covered with this kind of vegetation during the Ice Ages.

THE FATAL FLAMES OF AMBRONA

Enormous numbers of bones belonging to the extinct elephant *Elephas namadicus* were found some years ago at Ambrona, near Madrid. Half the bones of one huge bull elephant lay together, but the rest had been removed. Many other bones had been smashed open, presumably to extract the rich marrow. An elephant's skull had been similarly broken open, perhaps to remove the brains. In one place, large bones had been placed in a line across the ground, which was originally a bog, perhaps to form a causeway for men carrying away meat from the butchered carcase to eat on drier ground.

Other traces of human activity include stone tools and even wooden spears. But these men had not merely killed animals which happened to stray into the bog – the number of carcases was too great for that. Thin layers of burnt material have been discovered at the site, and this suggests that they had set fire to the grass and brush in order to drive the elephants into the mud, where they could easily be killed.

cattle. Hippos lived in England as recently as the Riss-Würm Interglacial period.

But, for these animals, the continent of Europe was a trap. To the north lay the ice, to the south the Mediterranean, barring their escape to warmer, southern lands. Each advancing Ice Age witnessed the extinction of more of these warmth-loving types of animal, while each interglacial saw a yet more impoverished fauna. Last of all came man, hunting the herbivores and competing with the carnivores for food and shelter.

Although the great climatic changes of the Pleistocene Epoch were undoubtedly responsible for the extinction of many species, it is now believed that in many cases quite different factors were involved. Some animals, such as the American *Glyptodon, Megatherium* and *Mastodon* had evolved before the Ice Ages and survived them. Others, such as the woolly mammoth, woolly rhino and steppe bison, had evolved as a result of adaptations to the Ice Ages. It seems strange that so many animals which had survived four Ice Ages and three interglacials should have found it impossible to tolerate the climate of the last interglacial period. The finger of suspicion has begun to point towards that potent predator, man himself.

It is significant that this wave of extinction affected mainly the large mammals and the large flightless birds. Still more suggestive is the fact that the extinctions did not take place simultaneously all over the world. In each continent they seem to correspond with the time of evolution, or the arrival, of races of man with advanced techniques of tool-making and presumably of hunting.

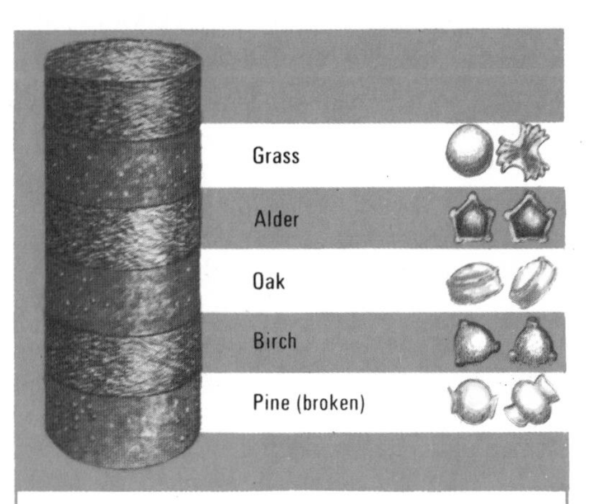

TELL-TALE GRAINS OF POLLEN

Each species of plant or tree has its own characteristically shaped pollen grain – pine, birch, oak, alder and grass, for example, differ widely in this respect. In the remote past, the hard outer coats of pollen grains were often preserved in mud deposits which were accumulating on the bottoms of lakes. If the climate changed, the vegetation changed too: new types of plant appeared and gradually became more common.

The pollen record of later times also demonstrates the gradually increasing influence of man on the environment. As he fells the forest and creates fields, so the pollen of man's crops and of their infesting weeds becomes more abundant. All of these changes are revealed in the above core sample. Pine and birch pollen indicate cold, dry conditions; oak and alder pollen at higher levels indicate that the climate was warming up, while the increase in grass pollen shows when man started to cut down the forests.

Nearly all mountains have their quota of beautiful blue gentians, and the same species often occur in the far north, although they do not occur anywhere in between. The plants are relicts of the Ice Age. They were once spread all round the fringes of the ice sheets, but they retreated with the glaciers and found refuge on scattered mountain tops or on the tundra.

Above: Mammoths were so common in parts of Europe and Asia during the Pleistocene that stone-age man even collected their bones to make huts. The hut shown here has been reconstructed from a circle of bones carefully excavated in the Ukraine.

A STICKY END

Between 15,000 and 14,000 years ago, thousands of animals met their deaths near what is today the centre of Los Angeles. They died, struggling vainly, in sticky pools of tar which formed where oil had seeped to the surface and evaporated. After rain, the pools might be covered by a sheet of water, concealing the fatal tar below. The bones of the trapped animals have recently been discovered, beautifully preserved, in the congealed and hardened tar deposits.

Two different types of animal are common in the tarpits. Some, such as the bison, horse and groundsloth, belong among the large herbivores which grazed over the surrounding plains. But they are far outnumbered by the carnivores which, attracted by their struggles, were in turn lured to their death. The remains of at least 150 bison have been found. The deposits also contain the bones of over 1500 dire-wolves and over 1000 sabre-toothed cats. Similarly, the remains of such birds of prey as the golden eagle greatly outnumber those of other types of bird.

Man in Ice-Age Europe

There was a rich fauna in Europe before the Ice Ages and it is very likely that man's immediate ancestors, such as Australopithecus (see page 151), lived and hunted there. But, apart from a single lower jaw, the earliest remains of man in Europe date only from a warm spell within the Mindel glaciation, some 400,000 years ago.

These earliest-known Europeans lived at a place now called Vértesszölös, near Budapest – or rather, they stayed there briefly from time to time, for they were probably nomadic hunters. A single skull fragment is the only human bone which has been found at Vértesszölös, but it is sufficient to show that these nomads were already in the process of evolving into true men of the species *Homo sapiens*. This find was accompanied by scattered traces of man's occupation – tools, the bones of the animals he killed, evidence of the use of fire, and even human footprints. Human remains of a rather similar type have been found elsewhere, dating from the Mindel-Riss interglacial, over 300,000 years ago. By this time, man had learned to make even more effective tools.

The Cave Men

It is not until more than 200,000 years later that we see our next glimpse of Ice Age man. By the time of the interglacial before the Würm glaciation, about 25,000 years ago, a very different type of human population had spread through Europe.

Below: A mammoth struggles in vain to get out of the sticky tar, watched eagerly by dire wolves, a sabre-toothed cat, and some circling vultures. They, too, will be trapped when they plunge in for the feast.

Neanderthal man, short in stature but powerfully built, had mastered new skills which allowed him to remain in Europe even when the ice returned during the last glaciation.

Ice Age Neanderthal man took refuge in caves or under overhanging rocks, where he could escape from the cold winds and build fires to keep himself warm. It seems likely that he used animal skins for clothing – a theory which has been confirmed by the presence of stone implements suitable for boring holes in skins and cutting them. He hunted the woolly mammoth, woolly rhino, deer, reindeer, ibex and chamois.

As the Würm glaciation progressed, the final evolutionary change in man's long history was gradually accomplished – he acquired the typically human powers of reasoning and imagination. By 35,000 years ago, Neanderthal man had disappeared and the earliest type of modern man, Cro-Magnon man, had taken his place in the caves and rock-shelters of Europe.

Above: One of the Los Angeles tar pits today.

The Ascent of Man

It used to be generally believed that human beings were very special creatures who had nothing in common with the animal world. Charles Darwin therefore caused quite a stir when he suggested that man is closely related to the apes; but few people would now deny that Darwin was right. Man certainly has much greater intelligence than any other animal, but he is biologically very similar to the other mammals and it is now generally accepted that he is a member of the group known as the primates. This group also includes the monkeys and apes, and it is clear that man is descended from some kind of ape-like creature. It is important to realize, however, that man is not descended from any kind of modern ape. Somewhere in the past there must have been an ape-like creature which was the ancestor both of man and of today's apes, but we shall probably have to go back at least 20 million years to find it.

Man's Primate Cousins

Living primates are divided into two major groups: the *prosimians*, and the *simians* or *anthropoids*. The prosimians are the more primitive of the two groups, and include the lemurs, the tarsiers, and the bushbabies. The anthropoids include the monkeys, the apes, and ourselves. The non-human primates are nearly all tree-dwelling animals. Their grasping hands and feet enable them to hold on to the branches safely, while their forward-looking eyes and relatively large brains allow them to judge distances accurately and to co-ordinate their movements as they leap from branch to branch. These same features have been very important in human evolution because, although we no longer live in the trees, we should not have progressed as far as we have without good hands and eyes and good brains to control them.

Dryopithecus, once known as Proconsul, was a Miocene ape which might have been an ancestor of the modern apes.

Opposite page: An African bushbaby—one of the few prosimians living outside Madagascar today. Its large, forward-looking eyes and its well developed 'hands' make it very efficient in the trees.

Left top: A male orang-utan shows the large, grasping hands which make him so adept at swinging through the trees.

Left centre: The ability to use his hands has been very important in both the physical and cultural evolution of man.

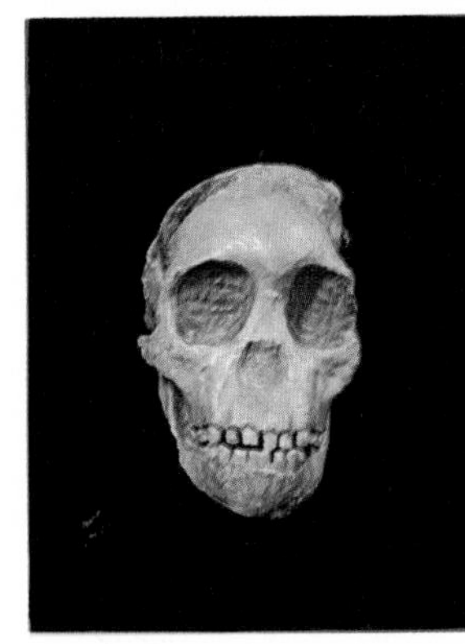

Above: The Taung skull, the first Australopithecine fossil to be found, was the skull of a child about six years old. The teeth were much more like those of a human child than those of a young ape, but the brain was much smaller.

The earliest primates appeared towards the end of the Cretaceous Period. They soon spread over all of North America and Europe, which were still linked together, and entered Africa a little later. Though monkeys succeeded in reaching isolated South America in the Early Oligocene, primates never reached Australia.

Even from their ealiest days, the primates were specialized to a completely arboreal life and probably descended only rarely to the ground. Their diet was mainly fruit, berries and leaves. Like many arboreal animals, they were probably more active at night. Life in the trees is very different from life on the ground, and demands very different adaptations. To climb and leap, an animal must be able to judge distance accurately. This is only possible if both eyes can focus on the same object. The eyes of the primates therefore gradually moved from the sides of the head to the front – the 'face'. This change was aided by the fact that the sense of smell became less used, so that the nose and snout became smaller. Primates also became different in another way. They took up a vertical pose, both while clinging and when leaping. This left the forelimbs free to become specialized for grasping and food-gathering.

The first undoubted monkey fossils come from Miocene rocks, but the animals must have been in existence long before that. They must have evolved from pro-

RAYMOND DART AND THE AUSTRALOPITHECINES

When the first Australopithecine or ape-man skull was unearthed at Taung in 1924 it was handed over to Raymond Dart, an Australian anatomist who was working in Johannesburg. Dart immediately saw that the skull bore resemblances to both apes and man and he was convinced that it belonged to an early member of the human line. He was supported in this by the famous palaeontologist Robert Broom (see page 95), but many other scientists disagreed. They said that as the skull was obviously of a young animal it might have become ape-like as it grew up. Dart was not to be put off, however, and he was later proved right. Many more fossil ape-men came to light in South Africa during the 1930s and 1940s, some of them discovered by Dart himself, and it became clear that the Australopithecines really were the fore-runners of man.

Dart also examined the antelope bones associated with the fossil ape-men, and noticed that certain bones were much more common than others. Lower jaws were particularly common, for example, while the lower ends of the limb bones were more common than the upper ends. Dart believed that these bones had been deliberately collected and fashioned for use as tools, and he described the ape-men as having a bone-tooth-horn culture or tool industry. But there is another possible explanation. Recent experimental work in South Africa suggests that these bones have survived merely because they are harder than the rest of the skeleton and less easily destroyed by weather or scavengers.

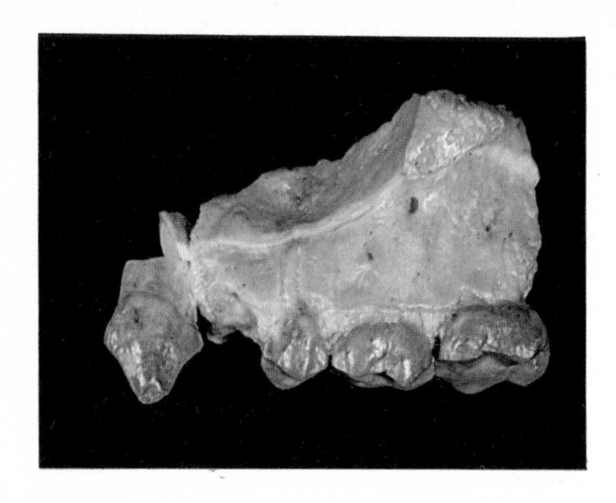

Left: *A fragment of the upper jaw of Ramapithecus, about 14 million years old. Although only a few jaw fragments are known, the teeth are so much more like those of men than those of apes, that we can be sure that the human, or hominid, line was already in existence.*

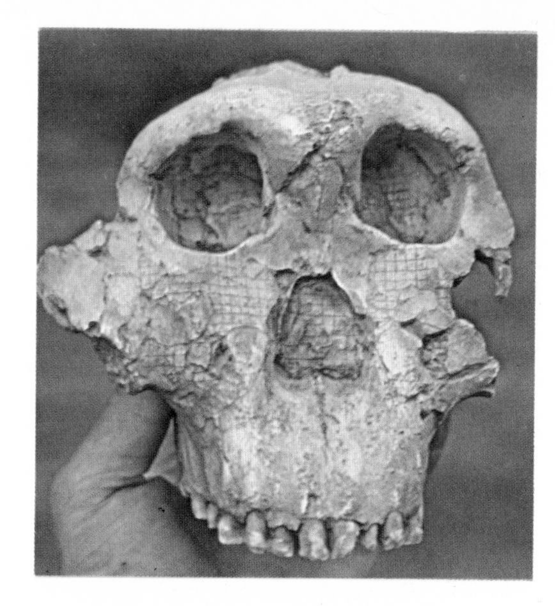

Right: *The skull of Australopithecus boisei, the large-toothed ape-man which the Leakeys named 'nutcracker man'.*

simian ancestors in Oligocene times, a development which spelt doom for most of the prosimians. There was a further reduction of the sense of smell in monkeys, resulting in a loss of the snout and a rounding of the face. Monkeys also had better brains than the prosimians, and were therefore more efficient animals. The prosimians could not compete with them, and they began to decline. Relatively few prosimians are alive today.

Living monkeys fall naturally into two main groups: the New World monkeys, with their flattened noses and sideways-pointing nostrils, and the Old World monkeys, whose nostrils point downwards. The two groups are not closely related, and may have evolved independently from separate prosimian ancestors. Little is known about the early monkeys but we can be fairly sure that it was one of the Old World group of monkeys that gave rise to the apes towards the end of the Oligocene. It was in Miocene times that the apes really came into their own. Large numbers of fossils have been found in East Africa and elsewhere; this suggests that numerous species ranged over the Old World about 20 million years ago. These Miocene apes are collectively known as *dryopithecines* and most of them were tree-living. Their teeth were rather different from those of the monkeys and it is thought that the animals fed mainly on fruit. The dryopithecines include the ancestors of all the modern apes as well as many extinct lines. It was originally thought that our own ancestors could be found among the dryopithecines as well, but it now seems likely that the human or hominid line branched off from the main ape line before the dryopithecines appeared.

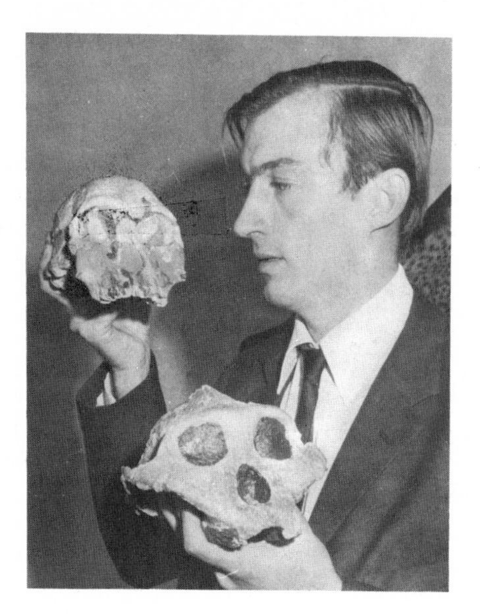

Richard Leakey holds the skulls of two hominids from Olduvai Gorge. The upper skull is that of 'handy man', while the other belonged to 'nutcracker man'. The latter was clearly the more robust of the two, but 'handy man' had a better brain and was able to make simple tools. 'Nutcracker man' was unable to compete with his more intelligent cousin and died out, but 'handy man' probably evolved into the first real man.

THE LEAKEYS AT OLDUVAI GORGE

Louis Leakey will go down in history as one of the most prominent and prolific finders of fossil men. When he died in 1972 he had spent the best part of twenty years unravelling man's history in the rocks of the Olduvai Gorge in Tanzania. Dr Leakey first started to explore the gorge in 1931, and he returned for short spells nearly every year, but it was not until the 1950s that he and his wife Mary began systematic excavations. They were later assisted by their son Richard and, between them, they have discovered dozens of fossil hominids. The various layers of the gorge, going back about 2 million years, have also yielded numerous primitive tools and the bones of hundreds of kinds of extinct animals.

The first hominid discovered by the Leakeys was the large-toothed ape-man *(Australopithecus boisei)* which they called 'nutcracker man'. But perhaps their most famous find was the smaller 'handy man', the earliest known tool-maker. He made the very simple stone axes known as Oldowan tools.

Louis Leakey was responsible for finding *Ramapithecus* fossils in Kenya, and he also discovered several true men *(Homo erectus)* at Olduvai and elsewhere. Richard Leakey has recently been finding fossil men in other parts of Africa and some of these discoveries may well help to fill in the many gaps that still remain in man's history.

The Parting of the Ways

The first real indications we have that man was in the making are some fragments of jaws from rocks in India and Kenya. They are about 14 million years old and the animals from which they came have been named *Ramapithecus*. Reconstruction of the jaws and face indicates that *Ramapithecus* had a head like a chimpanzee, but there was one very important difference: the form and arrangement of the teeth were much more like those of a human being than those of a chimpanzee. This does not, of course, mean that *Ramapithecus* was a man, but it does show that human evolution was well under way. *Ramapithecus* is now generally accepted as being one of our direct ancestors.

Working purely from the size of the face, we can estimate that *Ramapithecus* was a little smaller than a chimpanzee. We can only guess how he moved about because we have no bones from the limbs or body, but the hominids must have started to walk on two legs at about this time. Walking upright was a most important achievement in man's evolution, for it meant that the hands became free to carry and manipulate things. Our ancestors could thus begin to use simple tools. Later on they were able to make tools of their own. The upright stance also made the early hominids appear larger. This would have helped them in their hunting, for we can be sure that the animals were already varying their diet with meat. The changes in their teeth show that the diet must have changed.

Towards the end of Pliocene times the climate grew colder and the tropical forests in which our early ancestors had evolved began to shrink. Forests became savannah – a grassy habitat with scattered trees – but the bipedal apes were well able to cope with such a change. They evolved rapidly at this time, developing more and more human characteristics.

African Ape-Men

About three million years ago, South Africa was populated by creatures whose bodies possessed a remarkable combination of human and ape-like features. Remains of these ape-men were first discovered in 1924, when the skull of a young individual was found in a quarry at Taung, in Botswana. The skull was passed over to Professor Raymond Dart, who was immediately convinced that it belonged to the human line – although he actually named it *Australopithecus africanus*, meaning 'southern ape of Africa'. Many more remains have since been found, showing that the creature had an ape-like skull and face on top of a body which was much more like that of a man. It was in fact remarkably like the 'missing link' which Darwin had suggested.

Australopithecus africanus was about 120 centimetres (four feet) tall, and we know from the shape of his pelvic girdle that he walked more or less upright.

Excavating for fossil man in the Baringo district of northern Kenya. A single tooth, almost certainly belonging to a hominid, was found here in rocks about 6,500,000 years old. The site is thus very important, because we have no other hominid fossils of this age – nothing, in fact, between the 14 million year old Ramapithecus and the 3 million year old Australopithecines. Further excavation at this site may help to fill in this big gap in the story of fossil man.

The tooth arch of 'nutcracker man' (left) is much more like that of modern man than that of an ape (see below), showing that 'nutcracker man' was not just another ape.

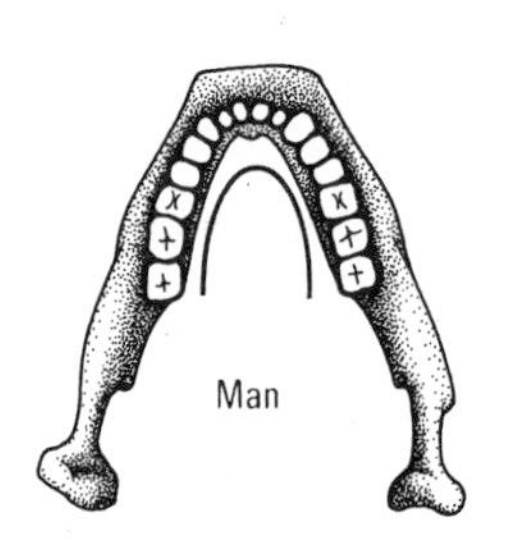

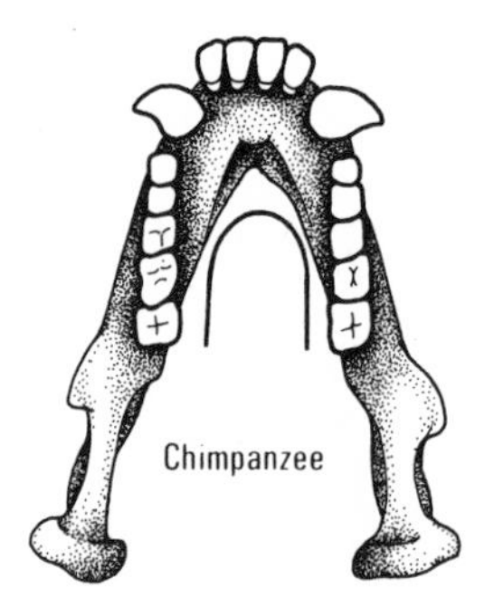

This was much more like that of modern man than that of the stooping or four-footed apes. More evidence for his upright posture comes from the skull, which was clearly balanced on top of the neck as our own skulls are. The skulls of apes project forward from the neck. The limb bones of *Australopithecus* were also more like those of man than those of the apes. The teeth were definitely of human type, but the jaws still projected strongly from the face and gave the animal a distinctly ape-like appearance. The rest of the skull was also ape-like and the brain was still very small – about 500 cubic centimetres in volume, compared with 1300 cubic centimetres or more for modern man.

Most of the fossil remains of *Australopithecus* have been found in cave deposits, but this does not necessarily mean that our ancestors were living in caves at the time. It is more likely that the ape-men merely sheltered in cave mouths from time to time and that some of them died there. Their remains were much more likely to be preserved in the limestone cave deposits than out on the savannah, and so these are the ones that are most likely to be found. The bones of baboons and various antelopes found with the ape-man remains suggest that *Australopithecus* was quite a proficient hunter. He probably used sticks and stones to kill his prey, and various animal bones for digging up edible roots; but there is no real evidence that these early ape-men actually *made* tools.

As more and more ape-man fossils came to light, it became obvious that *Australopithecus africanus* was not the only kind of ape-man living in South Africa three million years ago. Some of the fossils are of a much more heavily-built creature with larger jaws and teeth. This larger species has been named *Australopithecus robustus*. It lived in the same area as its lighter cousin, but its teeth suggest that it had a different diet and it may have lived in a slightly different habitat.

The similarity between the pelvic girdles of Australopithecus and modern man is strong evidence that the ape-men walked upright as we do.

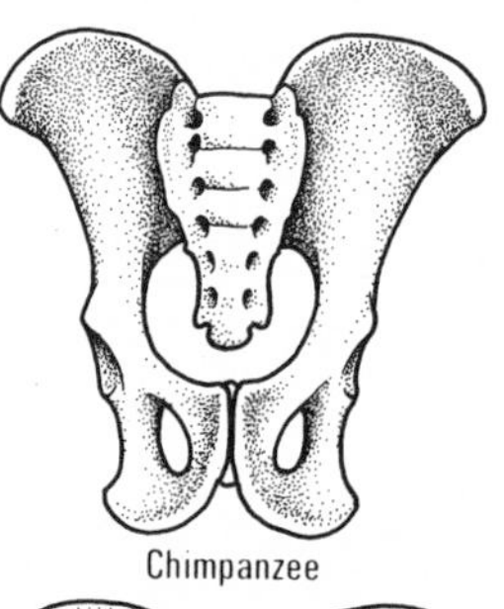

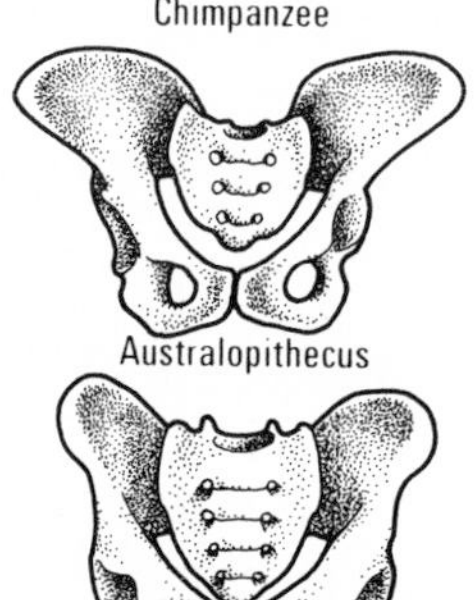

The Australopithecines, as these ape-men are generally known, must have been widely distributed in Africa in early Pleistocene times, and they probably occurred elsewhere as well. The Leakeys have discovered several specimens at Olduvai Gorge. Most of these are somewhat younger than the South African fossils, but there are still two distinct types. The more robust of the two has extremely large teeth and jaws and the Leakeys called him 'nutcracker man'. He is now known scientifically as *Australopithecus boisei*. Fossils of the more slender species date from about 1.75 million years ago and have been found in association with simple stone cutting tools. The creature presumably made these tools himself, and Dr Leakey named him 'handy man'. Because he made tools, he was thought to have been a true man, but most anthropologists now think that he was just an advanced ape-man. He had a bigger brain than the South African ape-men, but was otherwise very like them, and he is now called *Australopithecus habilis*. Although not a true man, he is thought to be our direct ancestor. His brain size continued to increase and, within a million years, he had given rise to creatures which were definitely men.

The First True Man

The descendants of 'handy man' evolved rapidly in early Pleistocene times and spread over much of the Old World. By one million years ago they had evolved into beings that were undoubtedly true men. They are known scientifically as *Homo erectus*, implying that they were much more like modern men *(Homo sapiens)* than the Australopithecines. The best known of these early men are Java Man and Peking Man.

Java Man was discovered near Trinil in eastern Java in 1891, when a Dutchman named Eugene Dubois found a fossilized

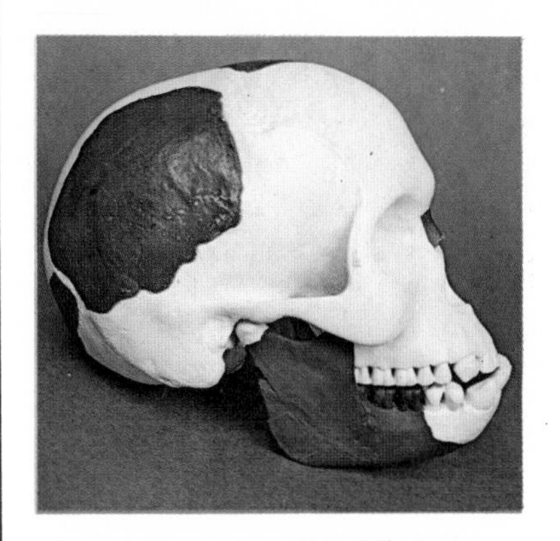

The reconstruction of the Piltdown 'skull' shows the actual bones in brown. The white parts show how anatomists restored the rest of the skull around the forgeries.

PILTDOWN – THE MAN WHO NEVER WAS

There was great excitement among palaeontologists in 1912 when it was announced that some pieces of a human skull had been found, together with a broken jaw-bone and a few teeth, near Piltdown in Sussex. More fragments were found later and, although many pieces were still missing, anatomists were able to reconstruct the complete skull. It had a brain case like that of modern man, but the jaws were distinctly ape-like and the canine teeth were large. The Australopithecines had not yet been discovered in Africa and Piltdown Man was acclaimed as the long-sought 'missing link' between man and the apes. He was thought to have lived in early Pleistocene times and he was named *Eoanthropus dawsoni* after Charles Dawson, the amateur archaeologist who found many of the bones.

When the Australopithecines came to light it became obvious that the large-brained Piltdown Man did not fit into the general scheme of human evolution. The scientists were puzzled and some wondered whether the Piltdown jaw actually belonged with the skull bones. In 1953 it was decided to subject all of the Piltdown bones to a thorough examination. The results were surprising. The fluorine content of the bones showed that they were only a few hundred years old, and chemical tests showed that they had been stained to make them appear much older. X-rays showed that the jaw-bone belonged to an orang-utan, and detailed microscopic study showed that the teeth had been filed down to make them look more human. Piltdown Man never existed. He was a very clever hoax. We don't know who played the joke, but modern methods of investigation will ensure that the palaeontologists are not fooled again.

skull. Further skulls and other bones were found later and we have a good idea of what Java Man looked like. With a height of about 150 centimetres (five feet), he was taller than 'handy man' and his hip and leg bones show that he had an even more upright posture. His face was much more human than that of 'handy man', although it still carried the heavy brow ridges characteristic of the apes. But the most important advances were in the brain. 'Handy man's' brain had a volume of about 600 cubic centimetres, but Java Man's brain was over 800 cubic centimetres in volume.

Peking Man, so named because most of his remains have been found near Peking, probably lived about 750,000 years ago. He was simply a later version of Java Man. His brain averaged about 1000 cubic centimetres in volume, but otherwise he was very similar to Java Man and it is clear that they were both forms of *Homo erectus*. Peking Man made simple stone tools, similar to those made by 'handy man'; he also used fire. His brain size suggests that he was becoming quite intelligent, and it is certain that he had already developed some form of language. He must have lived by hunting and by gathering various kinds of plant food.

Neanderthal – The First Wise Man

Homo erectus was widely distributed in Africa and Eurasia during the middle part of the Pleistocene. Fossils have been found in many places which show that

Part of the skull of Java Man (right), and a reconstruction of the skull of Peking Man (below). Although these early men had the heavy brow ridges and projecting jaws characteristic of the apes, their teeth were very like our own and their brains were much larger than those of any ape. Both belong to the species known as Homo erectus.

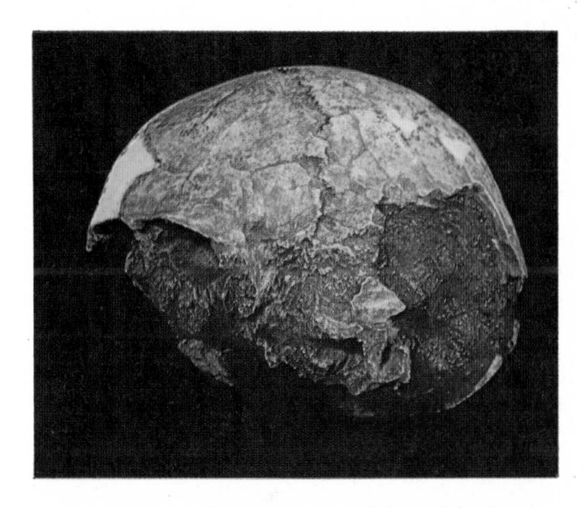

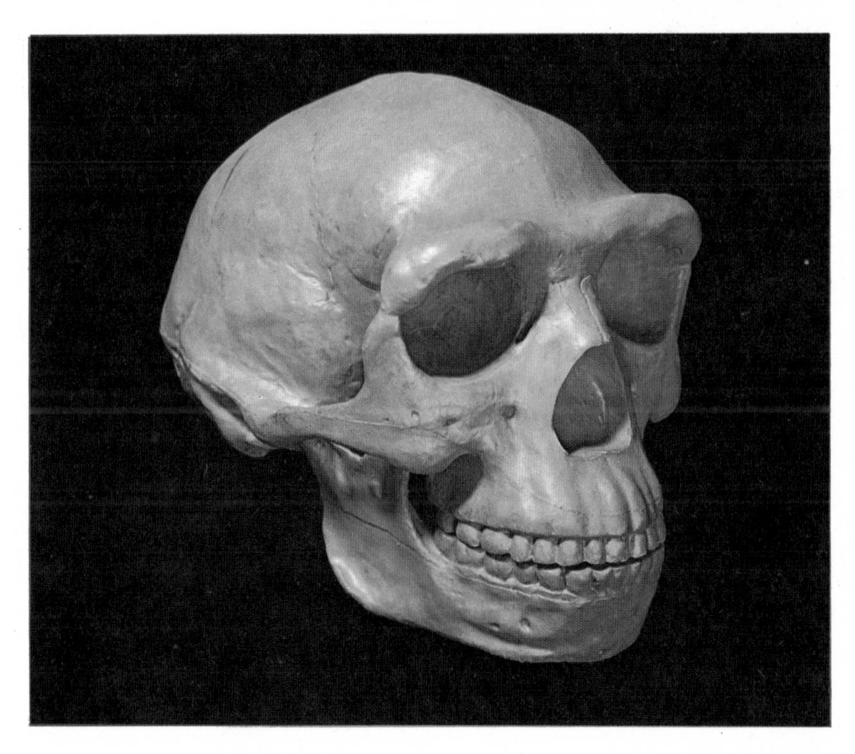

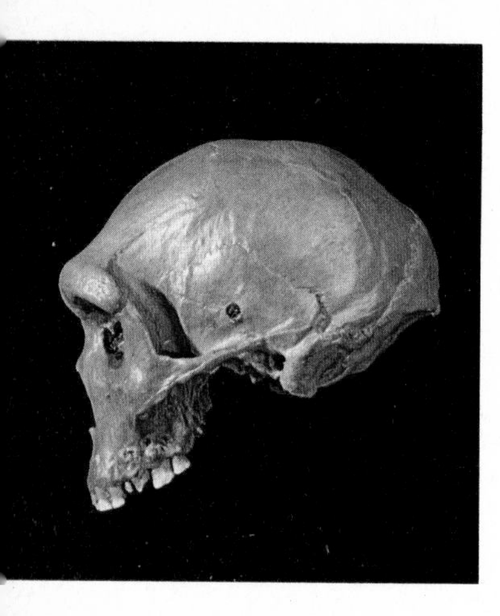

Above: The skull of Rhodesian man, a Neanderthal type who was living in Zambia (formerly Northern Rhodesia) about 50,000 years ago. Rhodesian man had enormous brow ridges and showed some other differences from the European Neanderthals.

the body gradually evolved until it was just like that of modern man. The brain also grew larger, and men living about 150,000 years ago had brains as big as ours are today. These were the Neanderthal men, so called because their remains were first discovered in the Neander Valley in Germany. With a brain and body like ours, these ancient men obviously belonged to our own species. *Homo sapiens*, or 'wise man', had therefore arrived. But the Neanderthals were still not exactly like us. They had rather heavy brow ridges and a receding forehead, and they also lacked a prominent chin. Although the Neanderthals have been found in Europe, they lived in many other places, including Africa.

The European Neanderthals had to put up with the rigours of the ice ages, but they managed very well by making their homes in caves and keeping themselves warm with fire. They also made simple clothes from the skins of the large animals which they killed for food. They made numerous hand axes and other stone tools.

The Arrival of Modern Man

About 40,000 years ago the European Neanderthals began to disappear and to be replaced by Cro-Magnon men. The latter were completely modern as far as their physical features were concerned,

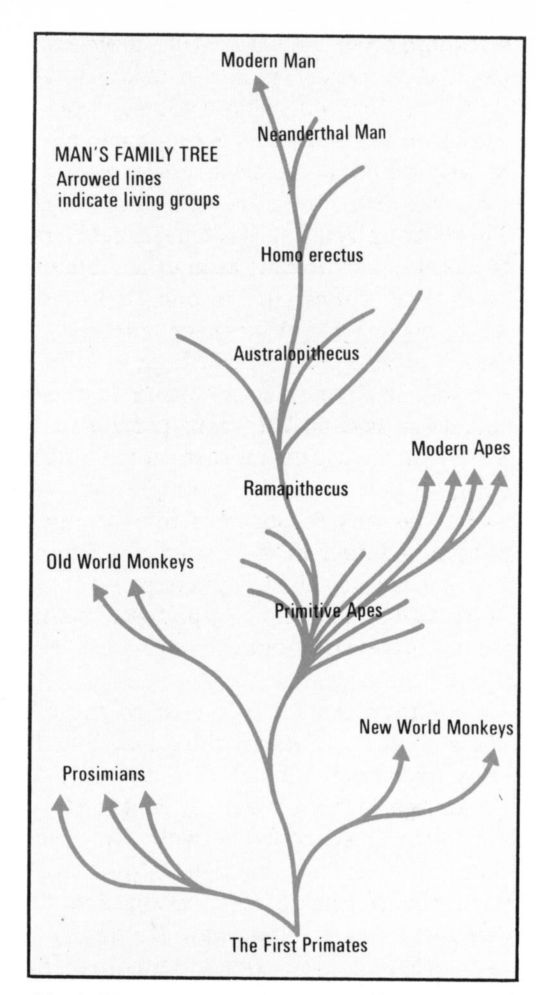

Man's family tree, showing his probable relationships with the apes and with the various kinds of fossil hominids.

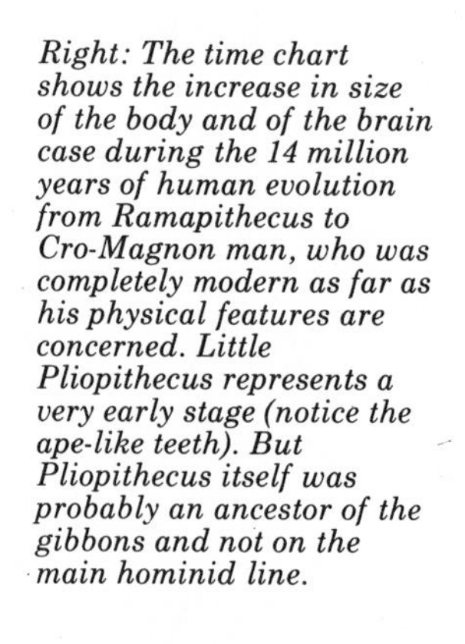

Right: The time chart shows the increase in size of the body and of the brain case during the 14 million years of human evolution from Ramapithecus to Cro-Magnon man, who was completely modern as far as his physical features are concerned. Little Pliopithecus represents a very early stage (notice the ape-like teeth). But Pliopithecus itself was probably an ancestor of the gibbons and not on the main hominid line.

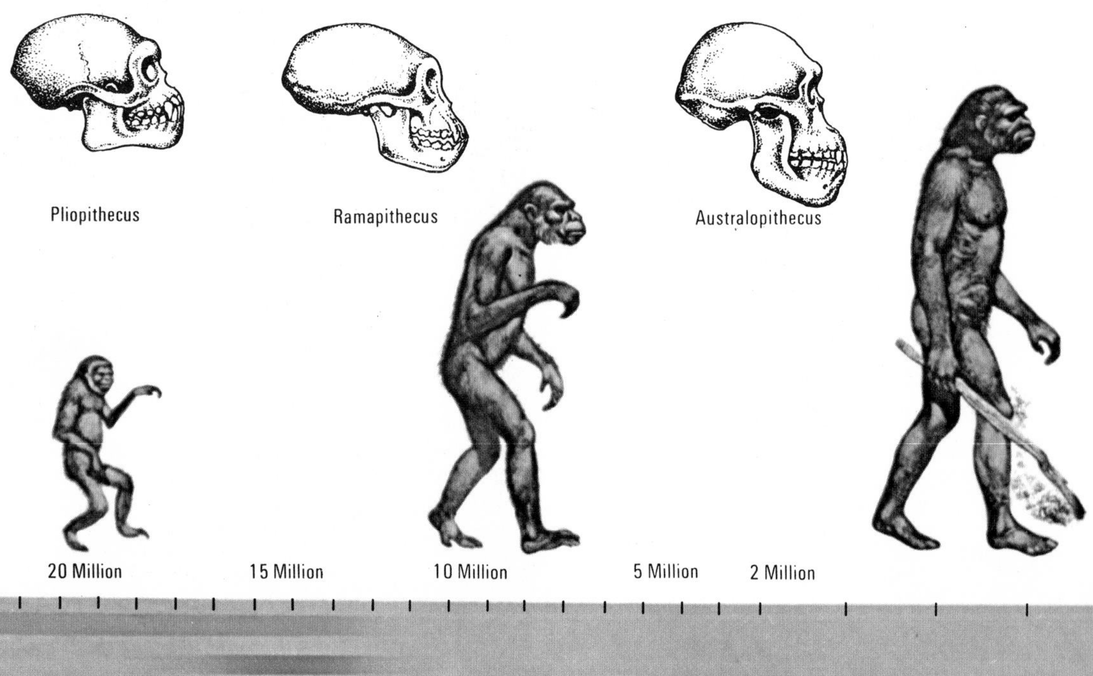

although they still led a very simple hunting and gathering existence. It is believed that the Cro-Magnons evolved from a group of Neanderthals living somewhere in western Asia and that they spread out from that area to populate the whole world. The various living races of mankind began to develop at this time as man established himself in different regions, but there have been very few other changes in man's body since Cro-Magnon times. Having acquired a good brain, man has spent the last 50,000 years putting it to use in improving his culture or way of life. Cro-Magnon man, for example, was an artist: he was responsible for the many paintings found in the caves of southern Europe. He also made rapid improvements in his stone tools and he was soon producing delicate drills and arrow heads from flints.

Cro-Magnon man belonged to the Old Stone Age or Palaeolithic, but his descendants soon moved into the Middle Stone Age with their improved stone tools and with bone tools as well. The New Stone Age came in about 10,000 years ago, with the Neolithic Revolution and the discovery of agriculture. Metals soon replaced stone and bone for tool-making, but agriculture remains the most important advance. It allowed man to settle down in one place and become civilized.

Above: One of the many fine cave paintings of Cro-Magnon man. These paintings tell us a good deal about the animals among which Cro-Magnon man lived.

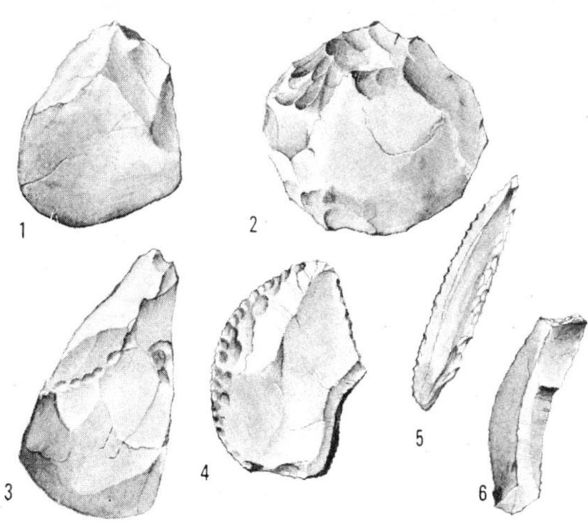

Left: Some tools made by our stone-age ancestors. 1. Oldowan hand axe of 'handy man'. 2. Chopping tool of Peking Man. 3. Hand axe of Homo erectus. 4. Hide-scraper of Neanderthal man. 5 and 6. Blade and scraper of Cro-Magnon man.

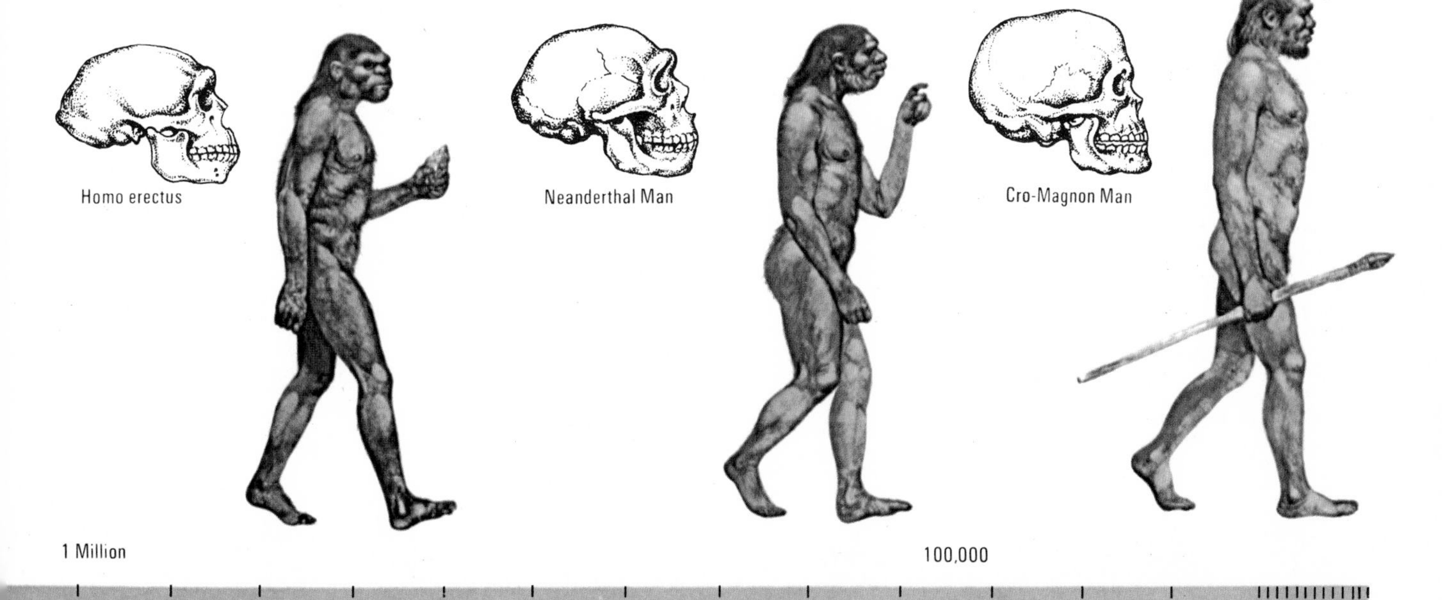

Index

Numerals in *italics* indicate an illustration

Picture Research: Penny Warn Jackie Newton

Acknowledgements
American Museum of Natural History 115; Associated Press 150 *centre right*; Australian News and Information Bureau 86, 134; Elso Barghoorn 41, 42 *top*; Biofotos/Heather Angel 28, 70; Martin Borland 32; British Tourist Association 149 *centre left*; J. Allan Cash/Susan G. Hill 99 *bottom*; Michael Chinery 17 *bottom*, 20 *centre left*, 24 *bottom*, 33 45 *bottom*, 46 *top left and inset bottom left*, 47 *inset bottom right*, 50 51, 59 *two bottom*, 65, 67, 68, 125, 145; E. N. K. Clarkson 61, 69; Cleveland Museum of Natural History 74; Dave Collins 98; Barry Cox 23, 95, 126; Michael Day 151; Diane Edwards 44, 45 *top*, 48, 83; Mary Evans Picture Library 9; Geological Museum London 34; Geoscience Features/ Basil Booth 8; Grønlands Geologiske Undersøgelse/D. Bridgewater 12 *bottom*, 13, 34 *top*; Hunterian Museum/Ian Rolfe 84 *bottom*; Imitor/Trustees of the British Museum 14 *top right and left*, 15, 17 *top*, 18 *top left*, 19 *top*, 21 *bottom*, 25, 37, 40 *top*, 43 *right*, 46 *inset bottom centre and right*, 47 *inset bottom left*, 52 *top left*, 82 *left*, 99 *top*, 103, 122, 124, 137, 149 *centre*, 150 *top left*, 153, 154, 155; Institute of Geological Sciences 64; Zophia Kielan-Jaworowska 20 *right*, 21 *top*, 116; R. Levi-Setti, Enrico Fermi Institute 62; Mansell Collection 12 *top*, 35, 106 *top*, 112 *right*; Pat Morris 1, 2, 5, 18 *top right*, 20 *bottom left*, 24 *top*, 27, 47 *inset bottom centre*, 56, 66, 73, 74 *top*, 75, 77, 79, 88, 106 *bottom*, 111, 118, 131, 147, 148, 150 *top right*, 152; Museum fur Naturkunde, Berlin 22, 23 *inset*, 120; Museum of Comparative Zoology, Harvard 93, 102; National Aeronautics and Space Administration 52 *right*; Natural History Photographic Agency, Brian Hawkes 20 *top left*, H. R. Allen (Courtesy of Royal Scottish Museum) 76, Ivan Polunin 80, Douglas Baglin 101, 128, 129 *left*, L. A. Williamson 129 *right*, Brian Hawkes 144; National Museum of Wales 46 *top left*; Novosti 14 *bottom left*, 146; Peabody Museum of Natural History 19 *bottom*, 104, 112 *left*; Princeton Museum of Natural History/Glenn Jepson 123; Royal Museum of Scotland 82 *right*; J. W. Schopf 40 *bottom*, 42 *bottom*; Sedgwick Museum, Cambridge 97 *right*; South African Tourist Office 156; Sternberg Memorial Museum 78; Swiss National Tourist Office 140, 143; M. W. F. Tweedie 29; United Press International 38; University of Miami/S. W. Fox 39; University of Witwartersrand 84 *top*; Zion National Park, Utah 59 *top*; Zoological Society of London 149 *top left*.